Answers to
Television Technology

An Encore

by

Cecil Smith

Newman-Smith Publishing Co., Inc.
Richardson, Texas

Newman-Smith Publishing Company, Inc.
2116 East Arapaho Rd., Suite 479
Richardson, Texas 75081 USA
(214) 231-6804

First Printing 1992
Library of Congress Catalog Card Number: 91-091439
ISBN 0-929549-03-1 (Soft Cover)
0-929549-06-6 (Hard Cover)

Printed in the United States of America.

Table of Contents

Computers 63

Recorders 85

Picture Displays 103

Part III: Systems

Appendix

Answers to Television Technology

Introduction

The questions appearing in *AV Video's* monthly "Technical Smithy" column have proven to be so popular that reprinting them in book form seemed a natural. There's been a crying need for straight answers in the muddled world of television technology. Artists have written for more information about the television canvas on which they "paint" their work. One-man shops have written to ask how to economically justify the purchase of new equipment. Technicians have asked how to go about troubleshooting equipment. Everybody wants to know about the latest "buzz words" bantered about in the industry. In your hands, you hold one man's attempt to cut through some of the haze.

As you read through the book, you'll notice that the style is not as formal and stuffy as other books about television technology. There's a deliberate reason for this—to try to personalize the scary, fast-paced world of television equipment. Too many people in the industry hide behind lofty production values and idealistic communications theories to avoid having to deal with the worlds of equipment, screwdrivers, mathematics (ugh!), and basic physical science. The conversational style of the questions and answers in this book is intended to alleviate some of the apprehension that we all feel when dealing in the world of television technology.

To try to make things easier, ***Answers to Television Technology*** has grouped the questions into three basic parts—"Equipment," "Signals," and "Systems." For "ease of read," the individual questions appearing in the magazine articles were categorized into these three areas. Each of the parts is further divided into sections covering specific topics (like "Cameras," "System Timing," etc.). Each of the parts begins with a "General Issues" section that serves as a gaggle of questions that refused to cooperate and be further categorized. If you're using ***Answers to Television Technology*** as a reference and want specific answers about a specific topic, see the index—it's as complete as possible.

Not all the questions appearing in all the sections apply to everybody. There are sooooo many market niches in this industry, that discussions about D-1 machines reads like ancient hieroglyphics to someone running a Super-VHS plant. If you are confused about an answer, it probably doesn't apply to your operations. Just in case, you may want to consult ***Mastering Television Technology: A Cure for the Common Video*** (also by Coleman Cecil Smith, P.E. and published by Newman-Smith Publishing Co., Inc.). There is an infinite number of levels of experience in this industry; use ***Answers to Television Technology*** to try to get one rung up.

Part I

Equipment

General Equipment Issues

Q: What is a "looping input?"

A: A looping input saps off a little bit of signal before sending the remaining signal on to be used by equipment that is connected downstream. A looping input usually has a nominal load value of 10,000 Ohms.

A looping input allows several pieces of equipment to sample the signal on a single cable with a minimum of distortion. The 10,000 Ohm looping input impedance allows several of these input loads to be connected in parallel on the line before the net impedance effect reaches the nominal 75 Ohm characteristic impedance of typical coaxial cable. (For more information about parallel resistance and impedance, consult any basic electronics text). Just remember the paramount rule–always terminate a source with the characteristic impedance of the system. For video circuits, this means that we must place a 75 Ohm resistor at the physical end of each cable.

Q: Why can't I bend my coaxial cable into a tight turn?

A: Coaxial cable is designed for very close tolerances between the center conductor and the outer shield. Try making a bend with an old toilet paper tube. Look at the way the tube deforms in the axis of the bend and the axis perpendicular to the bend. This is the same way that the shield of the coaxial cable reacts.

Since the distance between the center conductor of coaxial cable and the shield must be carefully controlled, the deformation created by a bend can have a very negative effect on the signal going through it. In the extreme case, it's like a garden hose that has been crimped to the point where no water will flow through.

I don't like to install cable with less than an eight-inch bend radius. If it must fit inside a space with a smaller bend radius, I use a right-angle connector that has been specially designed to maintain the distance between center conductor and shield.

Q: What's the best kind of connector to use–solder, twist-on, or crimp-on?

A: Hands down, the best technical way to install a connector is with solder. Solder provides a good electrical and mechanical bond between cable and connector.

But soldering has its price. It takes a while (about 5-10 minutes) for an experienced technician to install a solder-type connector. When you've got between fifty and five hundred connectors to install, that amounts to a lot of time spent simply installing connectors.

Enter the crimp-on connectors. Crimp-on connectors do not provide as good a mechanical and electrical bond, *but in most cases they will work just as well.* It takes about 1–2 minutes for an experienced technician to install a crimp-on connector. For the length of time that a particular television system will be connected in a particular configuration, crimp-on presents a cost-effective alternative to solder-on connectors. Remember, the configuration of today is probably not going to be the configuration tomorrow.

Then comes twist-on connectors. These connectors depend on the electrical bond that is created by forcing the cable conductors against the connector parts. The mechanical bond is provided by screw-thread cuts into the outer cable jacket. In my opinion, twist-on connectors should be used only in an emergency. Their mechanical and electrical integrity is such that they will fail with time and use.

For cables that will be used in field production operations, my choice is for soldered connectors regardless of cost. With all the coiling, uncoiling, twisting, connecting, disconnecting, etc. that goes on in field production activities, you need all the mechanical and electrical integrity you can muster. Even then, carry a spare–Murphy is alive and well!

Q: How does a "pixel" equate to "lines of resolution?"

A: There is no direct correlation between resolution measured in pixels (in computer parlance) and horizontal resolution measured in TV Lines (the gospel according to television). A <u>VERY COARSE APPROXIMATION</u> is:

2 pixels = 1 TV Line of resolution

Depending on what you are comparing, this formula may be way off. Pixels are frequently specified to encompass the totality of the viewing screen while TV Lines of resolution must consider the entire scan, including blanking. This "overhead" of unused pixels varies widely from equipment-to-equipment.

Q: What is a diode?

A: (Place tongue in cheek, there's not much truth to this explanation. On second thought, maybe there is!)

A diode slices, dices, and chops voltage. Available at your local electronics hardware store, diodes are frequently found in power supplies as the first active line of defense in the protection of fuses and circuit breakers. It is not unusual for a diode to die immediately before its fuse master dies. When a diode dies, the smoke contained in its body is frequently released into the atmosphere with a redolence that is reminiscent of a partially sated landfill.

Diodes are very intelligent. They can sense when the reliable operation of a piece of equipment is most important. When the critical time of operation is about to begin, they give up their life to make sure that the rest of the parts in the equipment are protected from too much use. This is why the symbol for a diode is drawn as an arrow running up against a brick wall (see Figure 1).

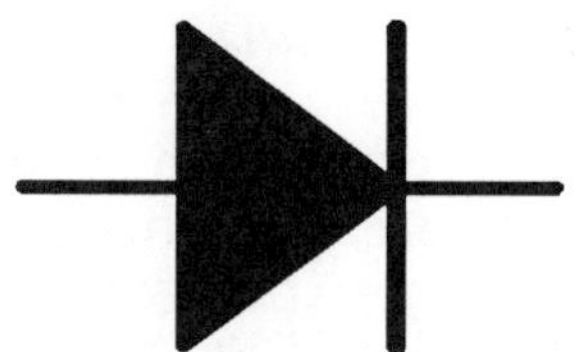

Figure 1
Schematic Symbol for a Diode

Q: What is TTL?

A: Depending on which side of the tracks you grew up, TTL can take on one of two meanings. For us guys that keep the train on the tracks, TTL means "Transistor-Transistor Logic." This is a quasi-standard method of dealing with digital control signals.

For the guys that like to play the train's whistle on their audio tape recorder, TTL means "Total Transport Logic." In their parlance, TTL means that the transport is smart enough to control the acceleration and deceleration of the tape so that the tape doesn't get stretched or broken.

For the guys that want to measure the Doppler effect on the whistle, TTL refers to a type of "video" output from acomputer.Computer CGA and EGA circuits output digital TTL signals. VGA circuits output analog signals that are incompatible with television equipment. These outputs are not compatible with normal television equipment and must be converted for television use. TTL usually cannot be used over a distance greater than about fifty feet unless signal repeaters are used.

Q: Why is gold a preferred electrical contact material?

A: Gold is frequently used in connectors to make good electrical contact because it is slow to oxidize and is an excellent conductor of electricity. If you look at some non-gold connector contacts, you may notice that the finish is dull. This is an oxide compound of the base metal of the connector. This oxide is formed by exposure of the metal to moisture and oxygen floating around in the air. Usually, this oxide is not a good conductor of electricity. You can imagine what happens when a bad conductor of electricity coats your connectors.

I'm not convinced that the extra costs associated with gold connectors are worth it in most operating environments. If the connectors are left connected for extended periods of time, the metal in the female connector protects the metal in the male connector from air (and vice versa), making the oxide take much longer to form. After the oxide does form, a pencil eraser works wonders. All bets are off on this discussion if you regularly operate in a caustic environment (like outside during an acid rain in Athens).

I can see no operational value in the gold connectors simply because they are better conductors of electricity. The amount of metal through which the electricity must pass is minimal, making the total resistance effect negligible in practice.

Gold or silver connectors *are* appropriate for portable operations, if you can afford it. Who knows where that cable has been?

Q: What is a transformer?

A: A transformer is a couple of coils of wire that are used to get electrons dizzy. Since the electrons must travel through the wire (kind of like these water parks where you go down the tubes) they get all caught up in the process of winding around and around and around in the coil of wire.

Now these electrons get so dizzy that they throw-up. But what they throw-up is not what you and I throw-up. Electrons throw-up magnetic fields. (I told you this was weird!)

When the electrons in the primary coil of wire throw-up onto the electrons in the secondary coil of wire, the secondary electrons get all upset. In their tizzy, they get higher and higher and higher until they just can't take this crazy ride anymore and must get out of the transformer. In all this madness, the exiting electrons have gotten themselves worked up into a much higher potential voltage state. Once recovered, these electrons with a higher voltage state can more easily do your work in a piece of equipment.

And *that's* the way it is. (Almost.)

Q: What is dealer markup in video equipment?

A: Usually, the dealer pays about 15%-20% off the list price for equipment. A profit margin of 3%-4% (12%-17% off list) is about the average sale price in Dallas nowadays. Mail-order "box" houses frequently operate with *very* slim margins in the 0.5%-1% range.

Do not expect to get equipment installed or serviced with these quoted margins. If you expect after-the-sale support or if you want the equipment installed, budget the full list price. Most dealers do not like to give away their expensive talents.

There are many, many variables. These tables are just rules-of-thumb that I use for estimating equipment budgets.

Caveat Emptor. ***Wait a minute – let me don my "resident cynic" hat.***

I keep seeing advertisements that make my blood crawl. No wonder this business is so confusing. NONSENSE appears in the ads and then you are expected to know what the nonsense really means. Here are some examples of bad, erroneous or misleading comments that drive me to drink. (At least it provides a good excuse!)

"... encodes at 5.5 MHz [sic]..."–I hope this isn't the frequency at which the chrominance is being encoded; that would make this box totally incompatible with the 3.58 MHz standard chrominance frequency in the USA. I've pondered long and hard about what the manufacturer is trying to claim and it could mean any one of Megathings (my own form of marketeerese). After considerable thought, *I think* that the 5.5 MHz is the clock rate at which the signals are being created (or sampled). Lots of smoke being blown here.

"... delivers full 500 line resolution ..."–Is there such a thing as *empty* 500 line resolution? Madison Avenue has had its influence on somebody.

"Even third generation tapes are broadcast quality."–This line comes from an ad for a product that has nothing whatsoever to do with video tapes!!!!!! *THIS IS B.S.*

"... is RS-170A approved ..."–How does a recommended standard (that's what the "RS" stands for) approve anything? It may meet the minimum performance criteria established in RS-170A, as claimed by the manufacturer, but who "approved" this product?

AARGH!

Q: If something breaks, what constitutes a warranty replacement?

A: I am not an attorney, but here's my opinion:

There are two issues here. First, was the part broken under conditions that are widely believed to be normal operational circumstances. Was the damage inflicted while the equipment was used in a way that was not usual and customary?

The second issue is the quality of workmanship and material. Has this same part failed repeatedly under similar operational circumstances? If it can be demonstrated that these failures are predicated on poor design, it is my opinion that the manufacturer should attempt to redesign the product for more reliable operation at his cost. The manufacturer has implied, from his specifications, that he has designed the product to withstand normal usage.

Q: Who makes the best equipment?

A: God.

Q: What procedures prolong battery life?

A: Don't expose the battery to extreme heat or extreme cold. If you're comfortable, it's comfortable.

When in storage, keep the battery charged. A battery will slowly discharge as it sits on a shelf.

Do not discharge the battery to the same point during each shoot. Completely discharge the battery (by continuing drain until equipment automatically shuts off) when time permits. Do not discharge the battery past the time when the equipment automatically shuts off.

Do not overcharge a battery. If your charger doesn't shut-off at the proper time, the battery could overheat and explode. Use only the type of charger recommended by the manufacturer.

If using lead-acid batteries (similar to those in cars, boats, and motorcycles) beware of hydrogen leaks. Remember the Hindenberg. Sealed cases have been known to unseal. Avoid arcing and sparking (in the technical sense, of course).

If the voltage from a battery drops, one of the cells has probably failed. Batteries are normally "constant current" devices–voltage changes as the battery is discharged. If the voltage drops, don't use the battery until the defective cell is replaced.

If a battery is dropped, examine for physical damage and monitor the output voltage.

If you want more information, Anton/Bauer (telephone /203/ 929-1100) has an excellent series of application notes on battery use and abuse.

Fractured Definition:

bat·tery \'bat-ə-rē\ *n* **1:** a cave near Gotham City **2:** repair procedure for a malfunctioning camera–*comp* ASSAULT **3:** a power source based on chemical reactions **4:** the first thing to fail on any important shoot **5:** object of attention when a cheap Producer says "charge it"

In response to a number of questions, clarification of the "battery" definition was necessary:

Some people were confused by "a cave outside Gotham City" in the definition of a "battery." To answer all the questions, I will ask a question. If nuns are in a nunnery, and armor is kept in an armory, then Batman would live in a . . . I really like this new jacket–novel idea to put the buckles in back!!

Q: Should NiCad batteries be stored charged or discharged?

A: I know it sounds like a political answer, but follow the manufacturer's recommendations for your particular batteries. In general, batteries should be stored in the charged mode. During extended storage, the batteries should be partially discharged, then slowly recharged (trickle charge) to full strength.

There are two enemies to batteries–heat and excessive discharge. As far as heat is concerned, keep your batteries as comfortable as you keep yourself. Remember that it's a lot hotter inside the battery than it is outside the battery.

As far a discharge is concerned, batteries slowly discharge even while they are sitting unused on your shelf. This discharge is OK unless the battery discharges to the point where the polarity reverses. If you store your battery in the charged state, the battery can discharge during storage for a longer period before damage occurs.

For those of you who need more details, Anton/Bauer (telephone /203/ 929-1100) has published a series of Technical Bulletins.

Q: Should I use Bakelite or metal jack frames?

A: In my opinion, I prefer the Bakelite jack panels instead of metal jack panels. Using Bakelite, an insulating material, the possibility of ground loops is considerably reduced.

There's a pretty good discussion about patch panels and ground loops in the front of the Trompeter Electronics (telephone /818/ 707-2020) catalog.

Q: What causes my editor (equipment-type, not personnel-type) to periodically go off and do its own thing?

A: Occasionally, an editor will stop in the middle of an edit, or refuse to enter the edit, or sit and stare back at you for no apparent reason. In my experience, this type of problem can be resolved by installing power line conditioning equipment.

Most editors are digitally controlled; i.e., they depend on minute, precisely-timed signal pulses for proper operation. If an errant pulse comes down the power line, it can easily interfere with proper operation of the editor. The time at which the errant pulse arrives determines how the editor reacts.

A power line conditioner reduces the number and strength of the errant pulses. Units are available for as little as $75 from computer supply stores. (Even Radio Shack has 'em!)

Q: What can I do if a piece of equipment does not perform to specifications?

A: I'm not a lawyer, but I am opinionated. In my opinion, the specification sheet is the manufacturer's promise of minimum performance for a particular piece of equipment. When a piece of equipment is purchased, performance to those published specifications becomes a condition of the purchasing contract.

Frequently, manufacturers publish their specifications months or even years before a product reaches the production phase. Sometimes the specifications contain a cop-out clause similar to "specifications subject to change without notice." This means, to me, that revised specifications may be created and publicly published to amend specified performance previously promised. This clause does not mean that the manufacturer has carte blanche to fraudulently represent the performance of his product.

The specifications must be complete. If the performance detailed in the specifications requires purchase of optional extra-cost equipment, it must clearly state that the optional equipment is required. The details on pricing must also detail those options. The legal point to this matter is that you, the consumer has a right to expect the level of performance promised by the manufacturer's specifications. If he can't meet the specifications, you usually have the legal right to find equipment that will meet the specifications *at no additional cost to you.*

In a disagreement with a manufacturer concerning failure to perform to specification, an attorney should be consulted early in the process. If your company has a purchasing agent, he must become the first person with whom you discuss the matter because he is the legal representative within your company. Each case is a little bit different, so *tread not with abandon.*

There are many manufacturers with whom I disagree with their business practices when it comes to applied technology. Manufacturers must assume as much responsibility for delivering promised products as their accounts receivable department assumes to secure payment.

Q: What are the "standard" tripod screw sizes?

A: The most common size for screws used in tripods is ¼"-20 although there are several other "standards." This standard means that the diameter of the screw is ¼" and that there are 20 threads-per-inch.

3/8"-16 is another tripod screw size that is popular.

Q: What's an EIA rack?

A: EIA stands for Electronic Industries Association. They issued a standard for the mounting of electronic equipment where the width of the front panel of a piece of equipment is just under 19". The height of the front panel will be sized in increments of "rack units" (RU) of 1¾". The front panel is wider than the chassis behind it.

In addition to the size of the panel, there is also a specification about the size and location of the mounting screws used to hold the front panel (and the chassis connected to the front panel). Although other screw sizes are sometimes encountered, the most common screw size is 10-32 (screw size number 10 with 32 threads-per-inch).

Frequently, the chassis is so heavy that "rack rails" are used to support the chassis and front panel instead of depending on the front panel screws. These rack rails connect to the front and back channels in the rack as well as to the equipment being supported. Frequently the rack rails are designed to slide the equipment out of the front of the rack for easy servicing.

For those of you who may be custom building wood consoles and racks, angle-iron assemblies pre-drilled for EIA rack mounting are available. I'm sure that there are other manufacturers, but I know that HOME (telephone /314/ 486-3111) makes 'em. If you do decide to roll your own racks and consoles, don't forget ventilation, power, and cable access for future equipment configurations!!! In most circumstances, it's far easier and cheaper to buy pre-fab racks and consoles. When you estimate the costs involved, don't forget to charge for your time for design, fabrication, and problem correction as well as for the materials used.

Q: What's the difference between "registration" and "convergence?"

A: "Registration" refers to the exact overlay of the separate red, green, and blue images in a color television *camera*. The image is focused by the lens onto the targets of the pick-up tubes or chips. To allow for the minute manufacturing and operating variations always encountered in electronics, the focused image is actually larger than the area that can be detected by a pick-up tube or chip. Each of the three pick-up tubes in a color television camera can be physically and electrically adjusted to detect the same portion of the focused image. The three pick-up chips in a solid-state imaging camera can only be physically adjusted to detect the same portion of the focused image. (Chip physical adjustment is normally performed only at the factory, leaving absolutely no physical or electrical adjustment that an operator must perform for optimization of registration in a solid-state camera.)

"Convergence" refers to the exact landing points of the electron beams in a picture tube (CRT–Cathode Ray Tube) in a color picture *monitor*. In a color CRT, there are three electron beams being generated. One electron beam is modulated by the red portion decoded from the video signal, one for the green portion, and one for blue. (All three electron beams may be created in one electron-shooting "gun.") On the faceplate of the CRT are three different kinds of materials coated on the inside of the clear glass. One kind of material glows red when it is activated by electrons, another material glows green, and a third glows blue. (Take a very close look at your picture tube–you'll see separate teenensy tiny red, green, and blue glowing areas.) Convergence adjusts the path of the electron beams so that the red-modulated electron beam activates only the red-glowing material, the green-modulated electron beam activates only the green-glowing material, and the blue-modulated electron beam activates only the blue-glowing material.

Registration and convergence errors can look *exactly* the same when viewing a color picture monitor. If the error changes with a change in cameras, the problem is probably registration in the color camera. (Confirm the error using the B-G and R-G display of the camera.) If the error doesn't change with a change in cameras, the problem is probably convergence in the color picture monitor.

Can a monochrome picture monitor have convergence error? No–only one electron beam and one glowing material.

Can a monochrome camera have registration error? No–there's only one pick-up tube.

Can a single-tube color camera have registration error? Well–maybe.

A single-tube color camera has to keep track of what color is being viewed at a given instant. (The pick-up tube is striped with tiny color filters.) If the single-tube color camera circuitry loses track of filter colors (by misadjustment or by component failure), the result can sometimes appear as a registration error.

Q: Why should I worry about keeping my equipment cool?

A: Keeping equipment cool is one of the commandments of the electronic industry. All those pieces of equipment in your racks are generating heat. Heat is a necessary by-product of making equipment operate properly.

Heat is also a dreaded enemy of electronic equipment. Excessive heat build-up in any given component within the equipment can cause the whole box to stop operating. (As we discussed a few installments ago, equipment operates on smoke. If you heat up an electronic component so much that it explodes and lets out the smoke, the equipment stops operating. Tongue is in cheek!)

Another type of problem occurs in diodes, transistors, and integrated circuits. There are teensy tiny wires connecting the outside leads of those components to the material within. When heated, these tiny wires become limber and expand (*anything* expands when heated). As long as the heat is applied and constant, there isn't any problem. The problem occurs when those tiny expanded wires are cooled excessively. Then they become brittle and attempt to contract. As they contract, they sometimes become so brittle that they break. *Viola,* instant equipment failure. Since heat is absolutely necessary for equipment operation, some type of constant cooling must be provided.

This is not to imply that equipment should be left on continuously. (Leave equipment continuously on only if it is necessary to maintain signal stability–a critical sync generator is the only piece of equipment that immediately comes to mind although there can be others in the typical corporate environment.)

So, keep the equipment cool. Don't block the air vents or fans of equipment with anything. This means top, bottom, front, back, and sides of equipment. The manufacturer of the equipment has tried to arrange the components in the equipment so that unblocked air flow will keep the operating environment of all those various components within acceptable temperature limits.

Keep the ambient temperature in the equipment room within the normal human comfort level, and possibly a little cool. (If you're comfortable, your equipment is probably comfortable.) There's no need to go into competition with Oscar Mayer.

If you can afford it, it sometimes helps to vent cool air into the top of a rack. (Remember that hot air rises and cool air falls.) To keep the air circulating in the rack, place the vent on either the left or right side of the top of the rack, not in the center of the top. (If you install the vent in the center of the top of the rack, sometimes the refrigerated air can become stratified and not cool the equipment in the bottom of the rack.)

The last control of the environmental temperature in which a component must operate is not so obvious. A piece of equipment naturally gathers dust. Charged dust particles are attracted to the charged electronic components. As the dust particles layer on a component, they trap an insulating air layer. After a while, the expensive cooling you are providing the equipment is not effective because each individual component has an insulating layer of air trapped between the heated component and the cool air. In short, cleanliness is again next to godliness when it comes to electronic equipment.

Q: What is a degausser?

A: It depends to whom you are talking–there are two types of equipment that share the same "degausser" term.

For the first several years I was in this business, the term "degausser" referred to a large coil of wire that was used to remove residual magnetic fields from a picture tube in a color picture monitor.

Any residual magnetic field changes the "purity" with which the tube displays a solid field of picture. If the picture is impure (has color tinges in areas of the picture–particularly at the corners) there is a problem with the electron beam conveying the one signal energizing the phosphors of the incorrect color. (For example, if the electron beam whose intensity is being changed by the red video signal is landing on some of the phosphors that glow green, there will be some areas of the picture that display a green or yellow tinge of color.)

The procedure for neutralizing any residual magnetic field resembles some sort of religious ritual. First you plug-in the degaussing coil and turn it on. Be sure that you are at least five feet from any picture tube (including the one you are attempting to adjust). Holding the coil so that the circle is perpendicular to the floor and parallel to the face of the tube you are trying to adjust, slowly approach the tube. As you approach the tube, move the coil in a vertical circle.

The colors in the picture will be going nuts!!! (At this point, it really impresses your boss if you are wearing a feather headdress, war paint, and can chant in a foreign tongue.)

Continue approaching until you are making a menacing circular motion about a foot from the face of the picture tube.

Stay that close for about fifteen seconds, then back away to the point where you can see no additional effect from the moving magnetic field from the degaussing coil. (Remember to continue the circular motion and all the magic chants throughout the process.) Then you can turn off the degaussing coil.

Some monitors and receivers have a built-in degaussing coil. This is simply a coil of wire that surrounds the face of the picture tube. Some sets have automatic cut-on and cut-off of the built-in coil. Other sets have manual cut-on and cut-off. Follow the manufacturer's recommendations for degaussing the particular monitor or receiver that you are using.

A monitor or receiver shouldn't require degaussing once the monitor or receiver is in place and has been degaussed. If you move the monitor around–well, that's a different story. It isn't unusual for the earth's magnetic field to create some problems that will require degaussing. I have a TV set on a Lazy Susan in my bedroom that changes purity depending on the direction that the set is facing.

The second piece of equipment that uses the term "degausser" is what we used to call a "bulk tape eraser." A bulk tape eraser is used to help the full-erase head in your VTR. The tape is exposed to the area of influence of the erase head for a very short period. To improve the completeness of erasure, the stronger magnetic field and exposure time provided by a degausser is much more effective.

The only common ground between the two uses of the term "degausser" is that they both are associated with creating a magnetic field. The differences involve how that magnetic field is used.

With this last comment in mind, I went to the ***Oxford English Dictionary*** to see what it had to say–here 'tis:

> **degausser:** To protect (a ship) against magnetic mines by encircling it with an electrically charged cable ... so as to demagnetize it. Also in extended use, to remove unwanted magnetism from, to demagnetize.

This is one device that can really blow your mind. (Ugh!)

Q: What service parts should I buy with new equipment?

A: With *any* piece of equipment, you should purchase the "Service Manual" for that piece of equipment. This is not the "Operator's Manual" that usually comes with the equipment, but is the publication with all the fold-out schematics and maintenance procedures.

Even though you may not be a technician, the "Service Manual" can really come in handy when you need to service the equipment. If your shop doesn't have a Manual, particularly if the unit is obsolete, your copy of the Manual may be the only available help. There are also some Manuals that contain adjustment and interconnect procedures that will allow you to get the most use from your equipment.

In the Service Manual will be details about what parts and equipment are needed to be effective in repair attempts. Some of this special equipment should be purchased, if they are not provided as a standard accessory of the equipment. Be on the lookout for special printed circuit card extenders, special jigs, and special test signals and charts. Again, even if you have no intention of personally attempting repair of equipment, the availability of the special repair parts often makes life considerably easier in a pinch.

Q: What should I look for when purchasing used video equipment? –Curtis Crowley, National Wildlife Fund, Washington, D.C.

A: That depends on the type of equipment. Many real bargains can be had out there. Many real dogs out there belong in a coral reef. I have paid as little as two cents on the dollar for used video equipment that fulfilled my needs. The problem is that it is always *caveat emptor*–let the buyer beware. You are probably not going to get any kind of warranty or guarantee once the equipment leaves the seller's premises.

If you know the operating and maintenance history of a particular piece of equipment, that information goes a long way in establishing the usefulness of the equipment if you purchase it. If the equipment had been removed from useful service, it probably will provide useful service for you. There may be some operational or technical quirks in the equipment, to be sure, but somebody in the past considered them acceptable.

Try to get a copy of any available documentation (operation manual, service manual, maintenance history, age, etc.). If the equipment is obsolete, this information may be crucial in continuing to maintain the equipment.

Open up the equipment and look inside. Look for an excessive amount of dust and dirt, frayed wires, and wires with brittle or broken insulation. Look for cracked circuit cards, charred or discolored components or areas of circuit cards, or any evidence of excessive heat. Look closely for any components that may be dangling and look as though they have been added to the equipment as an afterthought. Dangling parts indicate that the unit has been modified, either by a user *or* by the manufacturer.

If the equipment is no longer manufactured, open the equipment or look in the service manual to try to determine the number of generic and the number of manufacturer's proprietary parts used in the design. The greater the number of generic parts, the greater the potential useful life of the equipment. Beware of equipment from a manufacturer that has floated to the surface belly-up!! You probably can't get any support in finding proprietary spare parts.

Now for some specifics. Here are some of the things that would affect my buying price or decision to purchase a used piece of equipment:

COLOR CAMERA

Examine the quality of the lens and viewfinder. Are they securely attached? Remove the lens and gently shake it to feel if any of the glass lens elements have broken loose. Look for scratches in the front and rear glass lens elements. Turn the focus ring, the iris, and the zoom ring to feel how smooth their operation is–if the motion of any of them bind or feel gritty, the lens may need expensive service. Look for any dents in the case and operate all the switches.

Are the connectors on the camera damaged or worn? Plug in cables on the various connectors to make sure that they still work acceptably. Wiggle the connected cables to make sure that the connectors maintain connection. Watch for bent pins on multi-conductor cable. (Don't be too concerned about missing pins if the connector appears otherwise undamaged–some pins on many connectors are not used nor installed in the connector assembly.)

Look at the live picture (instead of color bars) produced by the camera. Zoom in and out to check for back focus. If the image goes badly out-of-focus as you zoom, the problem may be misadjustment of the macro adjustment of the lens. If the image goes only slightly out of back focus and it is a tube-type camera, an expensive back focus adjustment and registration procedure may be required.

Is there an excessive amount of noise (snow) in the picture? Are there any spots or dots (black, white, or color) present in the picture? For rapidly changing scene illumination or for moving objects in a scene, is there an excessive amount of lag or smear in camera response? Any of these problems may indicate that the expensive tubes or chips need to be replaced.

Leave the camera on with color bars for about a day and watch for excessive drift in hue. Look at all the color bars, but pay particular attention to the gray (leftmost) color bar. Although the problem may be caused by old capacitors that are moderately expensive to replace, a slow drift in hue normally indicates a design problem in the encoder circuitry and may be virtually impossible to correct.

While viewing a black grid on a white background, check the registration of overlay of the red, green, and blue imaging devices. If available, use the "minus green" ("-G") display in the viewfinder. Check the operation of the auto centering or auto shift circuitry. If there is any color fringing on the grid when using a chip-type camera, don't buy the camera–the expensive chip(s) and prism assembly probably need to be replaced. If there is any color fringing on the grid when using a tube-type camera (particularly in the corners), a moderately expensive adjustment of registration probably needs to be made.

Check the operation of manual and automatic white and black balance circuits, if available. Watch for any residual color tint near the black, gray, or white picture areas after the automatic circuits have been activated. Check colorimetry both with and without gain boost (0,+6, +9, +12, +18, etc.) If there is a residual tint, the pedestal, gamma, and gain circuits of the individual red, green, and blue imaging devices need to be adjusted (in a moderately expensive process).

If possible, have a technician check for proper operation of the sync generator circuit, as described later.

COLOR PICTURE MONITOR

Turn the brightness and contrast all the way clockwise while viewing a real scene. Watch for excessive blooming of the bright picture details. Watch for color fringes between adjacent scene details with large differences in brightness.

If available, display a dot pattern or a crosshatch pattern and examine the quality of convergence across the screen (particularly in the corners). If convergence is very far out of adjustment, a moderately costly adjustment of the monitor may be in order.

If possible, display color bars for about a week and watch for excessive drift in hue. Drift in hue usually indicates an expensive and time-consuming problem to fix.

WAVEFORM MONITOR OR VECTORSCOPE

Check the last time the waveform monitor was calibrated–there should be a sticker attached to the front.

Is the trace too dim to be usable, particularly in the 1 microsecond-per-division display of a waveform monitor? (This problem may be either an expensive aged display tube or an inexpensive misadjustment of an intensity limit control.) As you increase trace intensity, is the calibration of the trace maintained or does the trace change size?

VIDEO TAPE RECORDER

Exercise extreme caution on this one. Video tape recorders can become quite unstable with prolonged use and stable operation may be virtually impossible to attain.

Play back a tape and remove it to look for any damage to the edge of the tape that was caused by the machine being examined. If the edge is damaged, the machine needs an expensive realignment. Fast forward and rewind a tape from beginning-to-end; watch for speed variations and make sure that the end-of-tape detection circuits stop the tape at the appropriate times.

Play back an alignment tape and watch for any tracking problems or excessive noise (snow) in the picture. Tracking problems and excessive picture noise may be indicators of expensive video heads that need to be replaced, the need for an expensive alignment procedure, or the need for an inexpensive cleaning.

Listen to the playback audio for any changes in pitch in the sound. Changes in pitch may indicate either a moderately expensive problem with the capstan or capstan servo circuit.

Open the cover of the recorder and look for an excessive amount of dirt or oxide near the tape path. Feel the rubber pinch roller to see if it is hard or has flat spots or is out-of-round. Examine the video head drum assembly for any scoring or etching caused by tape. Examine the plastic parts for excessive wear.

Turn the VTR upside-down and look at the underside of the transport. If the VTR is driven with rubber belts, look at the condition of the belts. Expect to replace the belts at regular intervals to keep the machine in proper operating condition.

If the VTR is capable of editing, perform several edits using several different source tapes. If the VTR has a "framing servo," observe the length of time it takes for the light to illuminate after the machine is placed in the "play" or "record" mode–the longer the time to light, the less stable the VTR. Play back the edited tapes on another machine and watch for problems in the picture at the edit point.

TIME BASE CORRECTOR (TBC)

Connect it to the VTR creating the signal you want to correct and play back a tape. (The interconnection process will make sure that this particular TBC will interconnect properly with this VTR. Not all TBCs work with all VTRs.)

Play back a tape on the VTR and watch for excessive noise (snow) in the picture and for any residual instability in the picture from the time base corrected signal. If possible, watch the playback on a waveform monitor connected to external sync and look for an excessive amount of instability. (Some instability will always be there.)

Feel the top of the TBC for any excessive heat–the unit should be warm-to-hot, but not so hot that it burns.

CHARACTER GENERATOR

Check each mode of each key on the keyboard. Check each of the various fonts, font colors, and font sizes. When confirming operation, be sure to use a standard composite color video picture monitor–not an RGB monitor or a computer-type monitor. If possible record and playback the signal from the character generator to make sure you can get a good recordable signal out of the character generator.

Confirm the operation of any storage devices (floppy disk, hard disk, etc.). Confirm the operation of any special input device–watch for unstable operation of a digitizer tablet, light pen, or mouse. Confirm the operation of any built-in key circuit.

VIDEO PRODUCTION SWITCHER

Confirm the operation of each button or switch in each operational mode. Check the mechanical integrity of each pushbutton switch cap–they frequently become broken in ordinary use. Check the operation of any built-in color background generators or color bar generators–drift of hue or saturation in some designs has been a major problem (usually induced by a bad design) on some units. If the particular unit has an internal sync generator, evaluate that circuit as described in the next section. Confirm the operation of the blanking processor circuit, if the unit has one.

SYNC GENERATOR

You'll need test equipment and a trained mind to evaluate this one. Check the frequency of the internal master oscillator with a calibrated frequency counter–excessive drift or misadjustment of this circuit will induce massive problems in the entire system. Confirm the operation of the genlock circuit, if available. Confirm that the generated pulses meet broadcast standards. There's no need to accept less than broadcast standards here, considering the widespread availability and low cost of units that will perform acceptably.

Soapbox Time

We're knee-deep in the age of "guerilla television." Anyone who can afford any camera is a production company. As we get deeper into the cheap labor and equipment, it's going to become more evident that the ability to manipulate and interpret the technical television "canvas" on which we paint is what is going to separate the pros from the rest.

As an industry, we *must* meld the "engineering side" with the "creative side" to achieve the "communication side!" To do this, the "engineering side" must present its case in a manner that is no longer thoroughly detested by the "creative side."

Let's start a movement to break down all this technical stuff with its specialty language into something that people can use and understand. Eliminate the misleading marketing statements–they're starting to come back to haunt user and manufacturer alike. Eliminate bad technical information (there's a lot out there)–if you don't know something, say so (then go find the answer and report back). Work with your vendor so that you both learn about the technology instead of letting him dump his load of Bad Specifications (the technical acronym for this is "BS") on you.

Set aside a few hours each week to learn about another segment in your industry. If you're a technician, learn aboutproduction. If you're a producer, learn about the production. If you're a producer, learn about the technology. If you're a writer, learn about lighting. *Everything has to work together as a system!* The weak link(s) in our system can normally be found in the humans that are trying to get the electrons corralled into a good-looking picture.

Forget traditional titles and job responsibilities. If you're going to communicate your idea to your audience without distortion, you *must* consider everything, *including the technical.* If you're going to evoke an emotion from your audience by introducing a distortion, you *must* consider everything, *including the technical.*

In academia, the technology *must* be incorporated into the television production curricula. The "fun" communications curricula are *annually* producing 100,000+ highly educated "graduates" ready for this "exciting and high-paying industry" to vie for the 14,000 new jobs.[1] In the television segment of the communications industry, most of these jobs will be filled by graduates who know how to do something with the equipment, not those who just how to sit back and expound the wonders of the swish pan.

As an industry we must actively lobby the academic community to improve the quality of their graduates. Personally, I teach part-time at a local Junior College–what do you do to help the industry? We've all got something to contribute, even if it's just to be active on an industry advisory committee at the local school.

The only way our industry will survive the decade ahead is to pull our head from the sand and see *all* of the influences that are diluting the value of television as a communications and entertainment tool. The USA is the world's leading supplier of video programming, yet there is no significant presence of US companies manufacturing professional production equipment. There is no demonstrated need (other than economics and politics) to have improved television systems compatible with existing transmission systems, yet that is the dilution to which we have slumped as an industry. Advanced signal techniques (like digital tape) have been

[1]Bear's Guide to Earning Non-Traditional College Degrees.

around for almost two decades yet we are still yoked with new analog processing techniques that simply lengthen the time during which engineering costs can be amortized and a profit made.

What has made the USA unique is its ability to economically produce "high-quality" television programming. Other countries have attempted to penetrate the stronghold held by the USA, but none have been more than marginally effective. Now, we've got to make sure that other influences do not strip us of one of the few real markets we still control.

The economic environment is changing. Each of us must become more efficient. Each of us must produce as much material as we can of as high quality as we can. Each of us must develop a comprehensive overview of television as a communications and entertainment tool for our industry to stay alive for the next decade.

Q: How long should television equipment last?

A: This one's a lulu! It's full of ifs, ands and buts. But what the heck, I'll try almost anything once.

First, let's look at a couple of cycles through which television products go. The first cycle, the Product Development Cycle shown in Figure 2 determines how long a piece of equipment will be offered for sale on the market. Once the product is no longer offered, its value and technical capabilities begin to pale.

The Product Development Cycle basically starts with Research and Development (R&D). This is a search into finding a better way to do things. This the stage where basic scientific concepts are developed. R&D is the stage when CCD chips with imaging capabilities were developed. R&D is the stage currently involved in developing new recording media, among other challenges.

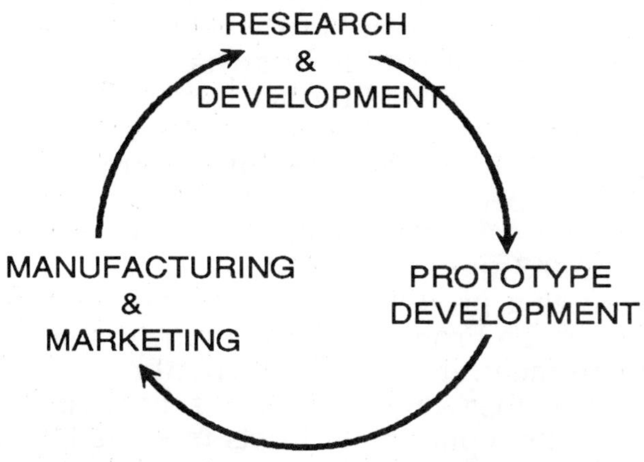

Figure 2
Product Development Cycle

The Product Development Cycle basically starts with Research and Development (R&D). This is a search into finding a better way to do things. This the stage where basic scientific concepts are developed. R&D is the stage when CCD chips with imaging capabilities were developed. R&D is the stage currently involved in developing new recording media, among other challengesThe Product Development Cycle basically starts with Research and Development (R&D). This is a search into finding a better way to do things. This the stage where basic scientific concepts are developed. R&D is the stage when CCD chips with imaging capabilities were developed. R&D is the stage currently involved in developing new recording media, among other challenges.

The stage of the Product Development Cycle where the R&D efforts are applied to a prospective product is what I call the Prototype Development stage. This is where a particular idea is molded into what is hoped to be marketable product. Here, sample units are built and tried. Here is also where particular features are added to make the product more salable. It is not unusual to see prototypes of products publicly shown to determine market reaction and to upstage introduction of competing products.

The last stage in the Product Development Cycle is the manufacturing and marketing of the actual product. At this time, almost all the decisions about features, price, and target market have been finalized although you will sometimes later see an "A" version with additional features (like the Sony DXC-M3 and DXC-M3A).

As soon as a product hits the streets, its replacement is probably in the Prototype Development stage of the cycle. Some manufacturers operate with two or more generations of prototypes in the wings. This means that a current product's replacement's replacement exists in some laboratory. Not every generation makes it to market, but the features and capabilities of some generations of a product are often incorporated into later generations.

What does this all mean? It means that there is a factor in determining the useful life of equipment that is controlled by the market. It means that there is an automatic mechanism that will obsolete equipment regardless of what you do. It means that a new, virgin piece of equipment sitting on a shelf has a limit on its useful life.

The elapsed time of one complete Product Development Cycle is dependent on the type of equipment. For example, I consider cameras to now be on a five-year cycle. I consider graphics generators to be on a three-year cycle. I consider recorders to be on a four-year cycle.

The duration of the cycles will change over time for several reasons. For example, manufacturer budgets for R&D have been significantly curtailed in recent years, making it take longer to develop the new ideas and technologies to be used in future products.

I have seen a marked slowdown in introduction of new television products in the last five years. Part of this slowdown is an indication of a saturated market, where everybody who can use the equipment has purchased it. Another part of the slowdown is caused by limitations of quality imposed by the standards within which equipment must operate (which is why the manufacturers are proposing the new "improved" HDTV standards and pumping most of their R&D dollars into it). Yet another part of the slowdown is imposed by the fact that we are just about using all the capability that silicon technology can give us–we've got to come up with some "new" technology that can give us greater capabilities in smaller packages with cheaper prices.

Now that we have considered the market forces behind equipment obsolescence, let's look at the cycle that limits the life of an individual piece of equipment. As shown in Figure 3, the minute that we buy a piece of equipment, we have established the starting point for wearing out a piece of equipment. Once we go through the "Equipment Purchase" stage and proceed through the "Normal Use/Routine Repair" stage, we normally reach a point where some type of "Major Repair" will be required to keep that piece of equipment in service. (I consider a Major Repair to be anything that costs more than 50% of the original price of the equipment.)

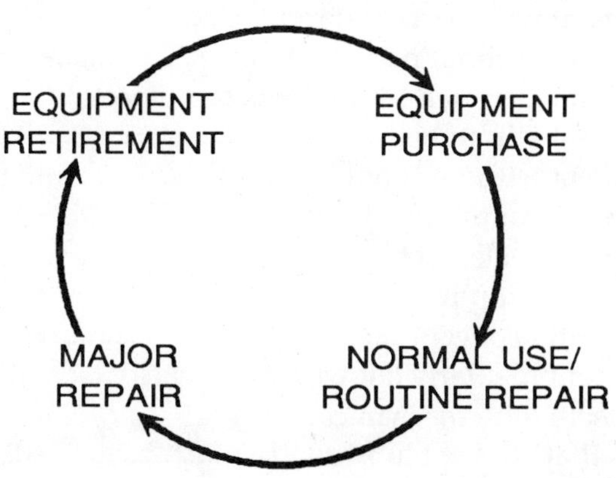

Figure 3
Product Useful Life Cycle

Like the Product Development Cycle, the Product Useful Life Cycle duration is dependent on the type of equipment and availability of parts. (Parts availability is determined by the popularity of the equipment, the length of time that a product has been discontinued, and the conscience of the manufacturer.)

Since an individual user cannot have any impact on the Product Development Cycle, the ability to lengthen or shorten the Product Useful Life Cycle of a piece of equipment becomes critical.

Let's take a look at some of the factors that impact the useful life of a piece of equipment. Let's, for the sake of argument, assume that *all* television equipment has an average useful life of 7.5 years. Now let's document what I think will be the impact of several variables associated with an individual piece of equipment to approximate the projected useful life. (The figures I'm citing are rules of thumb based on my experience and do not represent any scientific analysis.)

Factor	Impact
Equipment used in the field	-2.0 years
Equipment used only in a studio	+1.0 years
Equipment with parts that move	-1.5 years
Equipment without parts that move	+2.5 years
Equipment operated by students or novices	-1.5 years
Equipment with tubes	-2.0 years
Equipment without tubes	+3.0 years
Equipment continuously left on or "off"	+0.5 years
Equipment cycled between "on" and "off"	-0.5 years
Equipment supplied with conditioned power	+0.5 years
Equipment kept cool	+0.5 years
Equipment regularly allowed to overheat	-1.0 years
Equipment with mechanical parts that are kept lubricated	+0.5 years
Equipment kept clean	+0.5 years
Equipment that is not continuously "tweaked"	+0.5 years
Equipment with a plastic case or flexible chassis	-1.0 year

With my system, you take the average and add or subtract the appropriate adjustments to determine a projected useful life for an individual piece of equipment. You can develop your own average life and modifying factors based on your experience in your facility.

Q: What is the best way to be prepared for future developments in television technology?

A: It is my opinion that computers, digital communications, and television technology will merge over the next decade. To prepare for the transition, I would recommend:

1) Learn about conventional analog television technology. For the time being, the technologies associated with television operations are not going to radically change.

2) Learn about physics. I know, I know, UGHHH! It turns out that everything that we do as television communicators is based on an application of basic physical concepts. This portion of the training will require a basic understanding of mathematics–what better time to tackle it?

3) Learn about computers. DON'T go out and take a course in Basic or Fortran or Pascal or any other high-level programming language–unless your goal is to become a computer programmer. DO learn about basic operating systems and operating environments (for the Apple, Amiga, IBM, DEC VAX, etc.). DO learn about the generic applications programs (like word processing, spreadsheets, database, and graphics). As you master the applications programs, learn about the "MACRO" languages that are specific to a particular application–that's enough logical development in the programming environment to learn how to program any proprietary application (including robotics).

4) Learn about digital communications. Take a course about asynchronous communications and the use of "modems" for data transfer.

Q: I noticed a videographer open the top of a new VHS tape and stretch the tape a couple of times to make it really tight on the reels. I always though that a tape should be fast forwarded and rewound to pack the tape evenly on the reels. –Mike Costic, Video Taping Services, Cumberland, Rhode Island

A: That's a new one for me. I can't imagine why someone would want to do that to a poor little ole tape that never did anything to them. The repacking procedure that you talk about in your letter is the normal way of doing things. It breaks any residual binder ("glue") that may have squeezed out, removes any minor creases in the tape, and distributes any remaining dust or dirt in the cassette. It can mean the difference between a tape that tracks properly with a minimum of dropouts and a tape that is virtually unusable.

Unless this other guy had a lot of loose tape in the cassette housing, I guess he was trying to stretch the tape. Unfortunately I would think that all he would do would be to stretch the first few layers on the reel and grind the oxide into the plastic base of the adjacent layers. Unless he turns green when angry, there ain't no way he can have any effect on the layers near the hub. Anybody else got any ideas???

Prognostications.

Every so often, I'm asked to look into a crystal ball, past the swimming goldfish, and into the future. Here's the latest:

1. Congress will vote itself a pay raise. (Pretty safe bet!)
2. Four-chip color cameras will be marketed as the ultimate camera. The fourth chip will be unfiltered, providing luminance (Y) all by itself. The other chips will detect the Red, Green and Blue primary colors. This is the way the earliest tube-type cameras worked and, when properly registered, produced beautiful

pictures. Now that chips eliminate the registration requirements, we may see a technological full-circle.

3. Small-format digital video tape formats will be the norm by the magical 2000. At a conference in Montreux, Switzerland, Sony announced a digital format based on their BetaCam® transports. (DigiCam????) I expect a composite digital tape format using 8mm wide tape to appear next. (Digit8????)
4. Development of economically viable video disk technology will be deferred until the market is fairly well saturated with digital tape.
5. A satellite-based HDTV transmission system will be used to distribute theatrical releases and live events to neighborhood theatres. HDTV projection will be used to replace film projection. 70mm and larger film formats will proliferate to compete. HDTV will not be used in the home–a high-quality NTSC-compatible system will be used until a completely new digital transmission standard is developed in 15-20 years.
6. Solid-state larger-screen (19" and larger) color displays will not be commercially viable for at least 10 years. I believe they won't be viable until we start manufacturing chips under weightless conditions in outer space. The gigantic leap in yield from growing chips in space will make all kinds of stuff dirt (silicon) cheap.
7. At least 30% of all programming will be generated or visually manipulated by computer systems in 10 years. Parallel processing will place computers within the camera, recorder, transmitters and receivers. Full-scene animation of the old Disney quality will be "rediscovered" by cost-conscious producers.
7. I will be dead wrong at least once. (Probability of 100%. Not the weather bureau's probability, but the real world.) I'm always right except when I'm wrong–this man don't speak with fork-ed tongue!

Q: Can you offer any tips on the best way to troubleshoot equipment while using a schematic diagram? –Nancy Geisinger, KTXA-TV, Dallas

A: Nancy must be hard-up for another copy of my book. She was in the class I taught at Northlake College in Dallas, but it's a good question anyway. Hi, Nancy!

Before you open up the manual, find out as much as you can about the malfunction. You may want to separate your thinking into three areas–control, picture, and sound.

Control includes power problems and any physically identifiable malfunctions. This may include overheating, a VTR drum that is not locking, groans and moans emanating from the equipment, the smell of burnt insulation, etc.

Picture includes anything that is visible in the picture, on a waveform monitor, or on a vectorscope. Is it a problem with sync, luminance, or chrominance?

Sound includes anything that is audible from a speaker or that can be seen on a spectrum analyzer or oscilloscope.

Now open the manual and look at the Table of Contents and scan the first chapter or section of the manual. Look for any available troubleshooting charts. Some of the better manuals have decision trees that will lead you through common malfunctions to failed circuits or components. If you don't see any decision trees, go back to the Table of Contents to look for circuits and descriptions that match the areas of malfunction that you have identified. Now go to the section that describes the basic operation of the circuit that you consider the prime suspect and make sure that the circuit or system does what you think it does. If it couldn't possibly induce the malfunction that you observed, look at other descriptions.

Continue to narrow down the suspect circuits and mechanical systems.

Then, before I start really digging into the schematics, I look at the suspect circuit board or mechanical parts. Look for charred components, bend switches, seized parts, broken wires, loose screws (I know, I know ...), or anything else that may be amiss.

Finally, I'm down to the schematic. Find out which part(s) you suspect and get a magnifying glass to see what they look like. Look for cracks, cold solder joints, bad circuit traces, cracked boards, etc. If the problem is intermittent, I'd try some circuit coolant (Freon in a spray can) and a hair dryer to try to induce the problem at will.

If you've gotten this far without finding the problem, start using an oscilloscope to find out where the signal has strayed from specification. That ought to get you down to a handful of suspect components. Now start measuring the operating voltages applied to the pins of the active components (transistors, diodes, integrated circuits, etc.). If the voltages are not according to what the manufacturer recommends on the schematic, there's usually been a failure of resistors (open), capacitors (shorted), or diodes (shorted or open). If all the voltages are correct, you're basically down to active components and capacitors.

You can try the intellectual approach and get down to an individual component or you can shotgun new components into the circuit. Shotguning usually saves time for me, but it does cost additional money in parts.

In 25 words-or-less (± 12 dB), that's my basic approach to problem-solving at the component level. The key to the whole process is knowing where to look. That's the reason why the basic theory of how technical television works is essential to successful troubleshooting.

Q: What is meant by the term "clock" in my service manuals?

A: "Clock" comes from the digital side of the electronics family tree. The clock is the rate at which the digital bits of information appear and are processed. It's somewhat akin to a piano player using a metronome – each tick of the metronome (clock) tells the piano player (digital circuit) to do something.

In some applications, the clock serves as a traffic light to determine when one of several sharing devices gets a turn to use one wire for communications. This technique is used to "multiplex" several different signals onto one communications path. It's found all over the place, including inside computers.

In other applications, the clock serves as a "may I" command. Basically, the clock tells the digital signal to take one step forward. This is used in processing in a computer to move pieces of information into "registers" for manipulation. This is also used in the big lighted signs that have moving messages and drawings.

In any given piece of equipment, you may find several clocks for several different purposes. Sometimes they are synchronous (locked together in time). Sometimes they operate autonomously. Sometimes they operate autonomously when they are supposed to be synchronous, as the Space Shuttle technicians have painfully found out.

I guess that you can look at the clock signal as a digital biorhythm.

Q: Are X-rays a problem in video?

A: Yes, but not in the way you might think. X-rays, in the dosage found in airport X-ray machines are not a problem for video tape. (The metal detector, however, *may* erase the tape. X-rays *will* damage film.)

Where X-rays are a problem is in front of direct view picture tubes. Fortunately, this problem was recognized and corrected years ago.

When an electron is rapidly decelerated (as when the electrons in the electron beam hit the phosphors coated inside the picture tube), X-rays are normally generated. The harder that they hit, the stronger the X-rays generated. For the last couple of decades, all picture tubes in the U.S. must be impregnated with a lead shield so that X-rays at normal viewing distances are minimized.

I ran into an interesting situation a couple of years ago where some monitors destined for an overseas project were needed in the U.S. for system testing. We had to get a special permit from the government to bring the units into the U.S. and had to provide proof of when they were shipped out of the U.S. When you go overseas, there is no guarantee of the same protection that is available in the U.S.

Myths that Edith Hamilton never knew:

Digital is better than analog.

This should be rewritten that digital *can be* better than analog *if*

- You've got very deep pockets
- You need at least a dozen generations of tape for layering of scene elements
- You stay in the digital domain (digital switchers, digital processors, etc.) without ever going to the analog domain

Digital cameras are better than analog cameras.

There can be no such thing as a "digital" camera. If a camera were digital, it would produce only black or white, no grays or colors. Television was designed to operate in a manner that is **analog**ous to our eyes.

Despite the chest thumpings of marketeers, today's "digital" cameras are "digitally" controlled or process a "digital" representation of the original analog signal. In the future, we will see cameras with a digital signal output, but the imaging process itself *must be analog*!!!!

Solid-state "chip" imaging is better than tube imaging.

This should be rewritten that chip imaging *can be* better than tube imaging if

- There is very little light on the scene
- You're worried about power consumption or weight
- You don't want a "film look"
- You've got the money for the latest equipment

The latest version of hardware or software is the best.

This should be rewritten that the latest version of hardware or software *can be* the best if

- The new product has been thoroughly tested
- The new product has been subjected to rigorous quality control
- New technology, if used, has been successfully used in the field

The quality of product from professional equipment is no better than from consumer equipment.

This should be rewritten that the quality of product from professional equipment is no better than consumer equipment if

- You are viewing or listening to the product on equipment that does not have good resolution or distortion characteristics
- You are not subjecting the equipment to the rigors of everyday use
- You are not subjecting the signals in the product to the same number of generations of recording and processing that is required in professional operations
- You have relatively low expectations of product quality in professional applications

Q: Is there any use for old and worn 20-minute ¾" tapes?* –*Craig M. Jurkoic, Santa Barbara, CA

A: Leeetttsseeeee. I've seen the tape used for party decorations. NBC used to give away strips of tape to people who took their studio tour in beautiful downtown Burbank. The tape's too slick to be effective in the bathroom. It doesn't dissolve in the rain, so "papering" your neighbor's tree would not be a good idea. The housings can be given to kids for wholesale replacement of Lego blocks but unless the tape is removed, "Hang 'em High" will be replayed with the kid sister playing the lead.

Some people try to reuse the cassette housing on the pretense of saving money. Usually, the cheap repacking kits contain inferior tape and, considering the time involved, saves little (if any) money.

Cameras

Q: Should I disassemble a zoom lens to clean it?

A: Definitely not! Take it to someone who has already made a conscious decision to disassemble lenses as a career.

Q: Can a camera tube be burned even when turned off?

A: Yes. Even though the electron beam is turned off in the tube, the target material can still be seared by the bright light that has been concentrated by the lens. It is a good policy to always physically cap a camera to reduce the possibility of a permanent burn.

Q: How can I eliminate a burn in my camera?

A: In some instances, a burn can be reduced or eliminated by pointing the camera at an evenly—illuminated flat white surface and letting it sit. This procedure tends to equalize excessive charges that may be contained in the target material. After about six hours, if the burn is going to go away, it will begin to show. Contrary to some rumors, this procedure does not significantly alter the life of the pick-up tubes more than normal operation.

Another method that sometimes works is to place the tubes in a freezer overnight. Some people simply remove the camera lens and put the whole camera in the freezer, allowing plenty of room temperature warm-up before attempting operation. Nobody's been able to tell me why, but it sometimes works.

I have a theory why the freezer operation works. I think that the size of the crystalline structure of the target material is reduced when cooled. This reduction in structure size will encourage the discharge of any residual capacitive charges.

Q: How do I adjust the "CABLE LENGTH" control on my Camera Control Unit?

A: The CABLE LENGTH control adjusts for the loss of high-frequency video signal parts. These parts of the video signal consist of fine picture details and the color picture information. The easiest way to adjust the CAMERA LENGTH control is to look at the color bars generated in the Camera Head that is connected to the Camera Control Unit. Using a waveform monitor, the chrominance portion of the signal should peak at 100 IRE units in the yellow and cyan (second and third) color bars. If these color bars go above 100 IRE, there is too much of the high-frequency portion of the video signal. If these color bars do not reach 100 IRE, there is too little high-frequency signal.

Q: When looking at the picture from my camera, how can I tell where a spot of dust or dirt is located?

A: First, let's assume that the spot is dark. If the spot is in sharp focus, the dust is on the faceplate of the pick-up device. The spot may be cyan if the dust is on the red tube, magenta if the dust is on the green tube, or yellow if the dust is on the blue tube. You can tell which tube the dust is on if you look at the individual red, green and blue video outputs. It's unusual to see a colored spot, but still possible.

If the size and shape of the spot can be readily ascertained, but it is not in sharp focus, the dust is probably on the outside of the rear lens element or the color filter. Change filters to eliminate that location. Remove the lens to see if the problem is indeed there.

If the effects are very diffuse, the dust may be on the front of the lens. If the spot moves when you adjust the focus ring on the lens, the dust is probably on the outside of the front lens element or on a filter installed in front of the lens.

Do not under any circumstances disassemble a zoom lens to clean it!!!! You will soon realize that you can't make light of the bad situation (ugh). If the dust is inside the lens assembly, take it to a lens repair center. Your local neighborhood technician normally can't handle it.

Now let's assume that the spot in the picture appears white. Normally a white spot means TROUBLE (... that starts with T and rhymes with ... right here in River City). The spot is probably caused by movement of some of the dust left in the tube during manufacturing.

Try as they might, the manufacturers of pick-up tubes cannot maintain a perfectly dust-free environment. Some dust is present in almost every tube. If the dust starts moving around, it will probably start moving toward the target. If it ever gets to the target, it's probably there to stay. (Dust is frequently negatively charged and the target is normally positively charged.)

Depending on the chemical composition of the dust, the spot can really appear white, black or both. Again, the spot will normally be in pretty good, but not sharp focus. (The dust does not fall on the plane where the image is sharply focused, but behind it in the tube.)

What can you do about it? Retube.

POP QUIZ:

Define Plumbicon®:

a. a jail inmate who specializes in leaky faucets
b. an imaging tube developed by the Philips Corporation
c. a scam that is exactly vertical (as opposed to an out-of-plumbicon)
d. all of the above
e. none of the above

Define Saticon®:

a. someone who votes against an issue by sitting it out
b. a lazy jail inmate
c. the cousin of a corporate raider
d. an imaging tube developed by NHK.
e. all of the above

Q: How long should pick-up tubes last?

A: Now, in honor of this election year ... a pick-up tube should definitely last until it wears out. It all depends on how hard you push it in your operating environment. Generally most people see between 1500 and 3000 hours before they really need to re-tube.

You can tell when you need to retube when the pictures start to become laggy, increase in noise, lose sensitivity or lose sharpness.

Q: What is ginlock?

A: A secure liquor cabinet.

Q: Within a three-tube camera, what controls are sacred and should be touched only by a technician?

A: Of the many controls on a camera, the ones that need to be adjusted usually center on registration. In my opinion, any of the red and blue electronic controls for registration are fair game for most experienced operators. Keep those nasty little pinkies from any of the green controls and from any of the adjustments that physically adjust the red or green tubes.

RULES OF THUMB:

Don't touch any registration control that has the title "Green" associated with it.

Don't touch any registration control that has a mechanical lock associated with it (like "back focus" and "tube rotation.")

Don't touch any control until you understand what the name of the control implies. Vertical amplitude is the adjustment for the size of the picture on the vertical picture axis. Linearity is the adjustment for size of one-half of the picture relative to the size of the other half of the picture. Horizontal linearity adjusts this half-picture size relationship on the horizontal axis of the picture.

Keeping all this in mind, adjust the red channel registration controls to exactly overlay the green channel. Then adjust the blue channel registration controls to exactly overlay the blue channel.

Fractured Definition:

---

col·or bars \'kəl-ər 'bär\ *n* **1 :** drinking establishments that specialize in clean language (*antonym* off-color bars) **2 :** where certain pilots go to study hydraulic systems (hopefully *after* flying into the sunset) **3 :** a television test signal consisting of all possible combinations of full on/off red, green, and blue video signals **4 :** career goal of some interior decorators *v* writing graffiti with crayons on the walls of drinking establishments

Q: Should I white balance my camera first, then black balance, or vice versa?

A: The answer to this depends on how the circuitry of the camera was designed. Study the operating manual for your camera to determine the best procedure to use.

Q: How much of the picture should be white when I automatically adjust white balance?

A: This again depends on the design of the circuitry. Check your camera operator's manual.

Q: What is the "format" of a chip camera?

A: In this case, the term "format" has absolutely, positively *nothing* to do with video tape "format." Imaging "format" refers to the size of the sensed area of the focused image. As shown in Figure 4, the area scanned by the electron beam in a ½" pick-up tube is a rectangle (3 units high-by-4 units wide) with a diagonal measure of ½". A ⅔" CCD has a rectangular imaging area with a diagonal measure of ⅔". *In general*, the larger the "format," the higher the image resolution that can be sensed by the pick-up device.

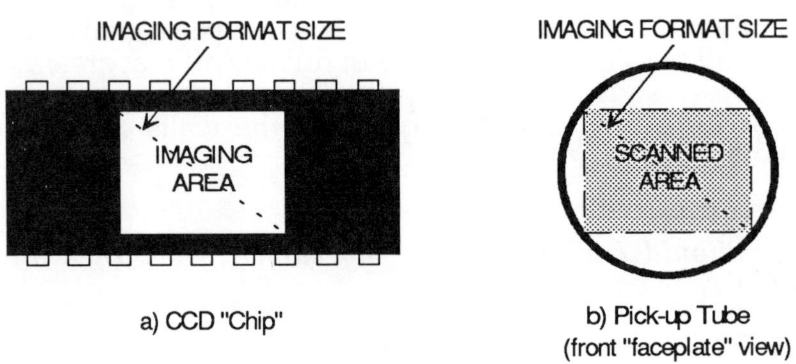

a) CCD "Chip"

b) Pick-up Tube
(front "faceplate" view)

Figure 4
Image Device "Format"

OK, all you math majors, if we have a rectangle with an aspect ratio of 3:4 and a diagonal measure of ½" within a circle (like the rectangular scanned area of a round faceplate of the pick-up tube), what is the minimum diameter of the circle? How about ⅔"? How about 1"? Try drawing it to scale if you can't visualize it.

With all this madness in mind, you can have ⅔" format CCD imaging devices that produce a video signal that is recorded on a video tape with a width of ½". The optical size of the sensed image area has nothing to do with the width of the tape onto which all the video, audio, time code, and control track signals can be recorded.

Now for one of my pet peeves. A "Camcorder" is a "Camera" and a "Recorder" in one box. If a live image looks good, it is the result of the "Camera" portion of the "Camcorder." If a played back tape looks good (regardless of the original source of the image), it is the result of the "Recorder" portion of the "Camcorder." There have been many times when I have heard about the superior specular highlight handling when looking at the live output from a particular E.I.E.I.O. video tape format "Camcorder." This is BUNK!!! Video tape formats have nothing to do with the imaging capability of the "Camera" portion of a "Camcorder!" Suspect the credibility of the source under those conditions! – Now back to Earth.

Q: What is auto iris?

A: Auto Iris is one of three methods used in television cameras to control the contrast (peak white brightness relative to peak black brightness) of the output video signal. Two other methods–Auto Target and Auto Gain–are sometimes used to control contrast, but rarely in ENG-style cameras.

Contrast is the visible effect of the ratio of the extremes of the video signal voltages. Remember that the pedestal voltage in a video signal corresponds to the brightness of the light reflected off a black cat in a coal mine at midnight. The maximum allowable voltage of a video signal corresponds to the brightness of the light reflected off a white cat in a snowstorm at noon. On a waveform monitor, the voltage for pedestal is displayed as 47.5 units above the voltage of the sync tip. The voltage for maximum white is displayed as 100 units above the sync tip.

Now that we have related "contrast" (the *visible* result) to "voltage" (the *electrical* parameter that conveys brightness information), let's design a circuit that looks at the voltage in the video output signal from a camera. As shown in Figure 5, if the video signal voltage is too low, the circuit will command some action to increase the signal level. If the video signal voltage is too high, our circuit will command some action to decrease the signal level. Our sampling circuit will rapidly and repeatedly sense the *average* voltage of our output signal.

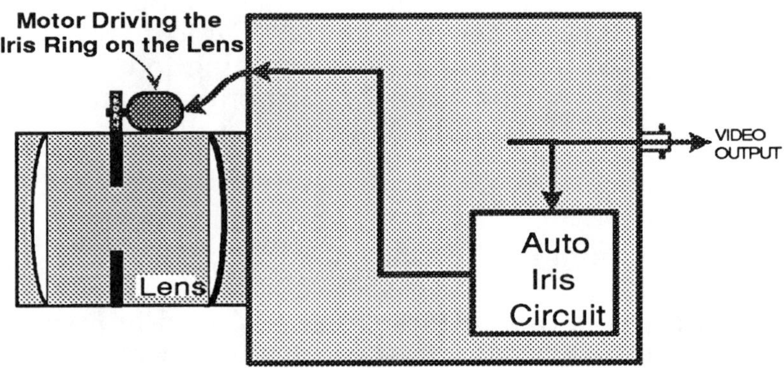

Figure 5
Typical Auto Iris Circuit

With Auto Iris, the output signal voltage is sensed and the iris of the lens is adjusted for the desired signal voltage. If the video signal voltage is incorrect, a motor turns the iris ring of the lens. When the video signal voltage is too low, the iris of the lens is opened up (to a lower f-stop number). When the video signal voltage is too high, the iris of the lens is closed down (to a higher f-stop number).

Now we've got Auto Iris. What's the down side?

Remember that when you change the iris setting of a lens,

you change the depth-of-field. (Depth-of-field is the *range* of distances in front of a lens when the image will be in sharp focus.) So Auto Iris operation changes the depth-of-field.

The other two methods that were mentioned, Auto Target and Auto Gain, introduce problems other than depth-of-field. Auto Target circuits adjust the operating condition of the pick-up tube. Auto Target circuits vary the lag (rate of response of the tube to rapidly varying light conditions) of the tube being controlled. Auto Target circuits usually are encountered in vidicon-tubed security cameras.

Auto Gain circuits control the output signal strength by adjusting the amount of amplification that the signal undergoes. It's kind of like a miniature "Gain Boost" switch (the one with +9, +12, +18, etc. on the side of the camera). As you increase gain, you increase noise, and *vice versa*. Auto Gain circuits aren't really popular in cameras anymore, with their modern day use relegated to processing and transmission equipment.

Q: How does auto focus work?

A: Not too well. I'm definitely *not* a fan of Auto Focus circuits. Unfortunately, like audio AGC (Automatic Gain Circuits), some of the prosumer equipment uses Auto Focus circuits. (In my opinion, the best adjustment for Auto Focus and audio automatic gain circuits is accomplished with a pair of wire cutters!!!)

Some cameras use infrared sensing of object distance from the lens. Basically, an infrared beam is transmitted from the camera into the scene. Depending on the angle of the beam reflected off the objects in the scene, a trigonometric calculation derives the distance of the object from the lens. Once derived, the calculated distance is used as a reference in the command of a motor attached to the focus ring of the lens. The focus adjustment is made by a dumb motor driven by dumb electronics instead of a cameraman's grubby paws driven by a (relatively) high degree of intelligence.

Other cameras use ultrasonic sensing of object distance. It's like the sonar in ***Run Silent, Run Deep***–just listen for the echo. The echo is inaudible to human ears because of its high frequency–but, just hold onto the dawg! Circuits are used to determine the distance to objects by the measuring the amount of time delay from the transmission of the pulse of sound until the receipt of the sound echo. Other circuits command the motorized adjustment of the focus ring on the lens.

Other cameras use two mirrors (yes, it *is* done with mirrors) to determine distance through triangulation. Basically they're looking for the same brightness information from each of the two mirrors. UGH!

The remaining method is probably the best working of Auto Focus methods. It looks for the high-frequency components of the video signal and uses a microprocessor (an itsy bitsy computer circuit) to command the focusing of the image. When the image is in focus, the high-frequency video signal components are maximized.

Even with the careful control of all the factors that can fool auto focus systems, there is *always* going to be confusion about which are the desired objects in a scene and where they exist. For example, the fence posts may be in focus, but your kid riding the pony in the background can't be seen. How's a dumb camera to know?

If you want to know more about auto focus systems, the best discussion I have seen is in ***Handbook of Video Camera Servicing and Troubleshooting Techniques*** by Frank Heverly and published by Prentice-Hall in 1986. This book is a folksy discussion about servicing consumer and prosumer equipment.

Fractured Definitions:

dif·fer·en·tial gain \dif-ə-'ren-chəl 'gān\ *n* **1** : change between the amount of power from the engine to the amount of power applied to the wheels **2** : change in weight before and after a diet **3** : nonlinear video distortion where a change in luminance amplitude induces a change in chrominance amplitude

dif·fer·en·tial phase \dif-ə-'ren-chəl 'fāz\ **1** : change from full moon to new moon change from young adult to middle age **2** : nonlinear video distortion where a change in luminance amplitude induces a change in chrominance phase

HAD \had\ **1** : once possessed **2** : taken advantage of ("I've been HAD.") **3**: hole accumulator diode

M·O·S \'əm ō əs\ **1** : what's not Larry's, Curley's or Shep's (it's Moe's–get it?) **2** : without sound **3** : metal oxide semiconductor

Do you think it could be that my hat's too tight?

Q: What is a faceplate?

A: The faceplate is the glass front of a pick-up tube through which the image is focused on the target material. The light-sensitive "target" material is coated on the inside of the faceplate. (In actuality, a "transparent electrode" material is between the faceplate and the target material. This transparent coating conducts the current which will be used to develop the video signal. The coating is connected to the conducting ring surrounding the faceplate on the outside of the tube.)

Q: What are "zebra bars" and what are they used for? – Richard Scott, Menorah Productions, Canby, California

A: The Zebra Bars in the viewfinder of a camera indicate when there is too much signal being created in a particular area of the picture. Where you see the zebra are those picture areas where there will be little or no detail because they will be eliminated in the "white clip" circuit of the camera or of the VTR. The zebra pattern is the way a camera warns you that this loss of picture detail will probably occur.

To understand the process, you must keep in mind that when a piece of electronic equipment is designed, the limits of voltage (or power or current) within which the equipment will operate properly are established. This is true of all equipment, electronic as well as mechanical. We shouldn't impose too great a load or the equipment will break down.

In video, a standard has been universally agreed that the signal voltage level should never be greater than 1 Volt (sync tip -to- peak white). If white picture details have a signal voltage greater than 1 Volt above sync tip, it can present problems in the proper operation of equipment. In a VTR, there can be severe problems with the color and black-and-white portions of the same signal interfering with each other. In a transmitter or modulator, too strong an incoming signal may create interference with adjacent television channels. In a switcher, the signal may "cross-talk" with other signals.

Sooo, in all these pieces of equipment there is usually a "white clip" circuit that acts like hedge clippers to lop off those portions of the video signal that may create massive operational problems. (The designer figures that you'd like *some* signal, albeit distorted, rather than *no* usable signal.) If you look at a clipped signal, the creases and folds in a white shirt will disappear; glare on faces will remove or change color; and colors in bright picture areas will appear pasty.

Here's an easy analogy–I like to call it "toilet training." You can think of the electrons in the video signal equivalent to water. The maximum amount of signal that the equipment will accept is the size of the tank of your toilet. What would happen if you tried to put too many molecules of water (electrons) into the tank (maximum signal)? Get a mop!

Now let's design a float that will stop the flow of water before we make a real mess. (Take a look inside your "loo" as you flush and watch how it works. A white clip circuit works the same way. I'm having a tough time at this point not saying something about the material being conveyed by both the video signal and the toilet, but I'll be nice.) *That's* your "white clip" circuit. You know that things are OK when you hear that the water has stopped flowing into the toilet and that there is no mess. The Zebra Pattern is your indication that the white clip circuit is operating (you can't hear the flow of electricity stopping and you sometimes can't immediately see the mess).

You need to know when the white clip circuit is operating so that you can optimize the desired white picture areas. If you see a zebra in areas of the picture where you want to see details, reduce the signal voltage being created by the camera. You can reduce the voltage by reducing light on the object, closing-down the iris (increasing the f-stop number), or switching to "0" gain boost (instead of "+9" or "+18").

***Q:** Why does a chip video camera have a shutter and a tube camera doesn't?" –Steve Nichols, Camera One Video, Inc., St. Paul, Minnesota*

A: Either of the two types of cameras *can* be equipped with a shutter.

Cameras are equipped with shutters for two basic reasons. First, a shutter can be used to create a sharp image of a moving object. This is similar to the technique used in 35mm still photography where you want each individual frame to have the sharpest possible picture.

If, while watching a film or tape at normal speed, you freeze a frame of a scene, you will notice that any moving objects in the scene are blurred. (Go ahead, try it on your VCR, if you've got still frame.) Moving objects in scenes are allowed to be blurry because the eye is relatively insensitive to resolution (sharpness) of a moving image. (This is why some images–like sleek sports cars–are often shot with a blurred trailing image. The blur conveys motion because that's the way that the eye is accustomed to seeing images in motion.)

When a camera with a shutter is adjusted to reduce the amount of time during which the image is detected, moving objects become sharper in each frame. When the individual frames are sharp, the pictures seen in slow motion or in still frame will be sharp. Uses for extremely short exposure times include sports analysis, scientific research, missile tracking–almost any application needing good visual resolution of moving objects.

Normally, each North American television frame is exposed for about $1/45$ second. (This is the mean between the $1/30$ second frame rate and the $1/60$ second field rate. For European-developed systems, the mean becomes $2/75$ second.) With a shutter, the exposure time can be reduced to $1/500$, $1/1000$, $1/2000$, or even $1/10000$ second. You can never get an exposure time longer than the field rate of the television system being used.

The second reason for equipping a camera with a shutter is the ability to vary the depth-of-field of the scene viewed by the lens. (Depth-of-field is the range of distances in front of the lens where objects are in sharp focus.) By reducing the amount of time during which the scene is detected, the depth-of-field can be reduced (because the lens iris must open-up to compensate for the reduced time during which the light can accumulate on the imaging device in the camera). Conversely, in order to use a short exposure time to sharpen each television frame, the scene must either be lit with extra lighting or you must put up with a reduced depth-of-field.

One thing to point out here–when you set the shutter on your television camera to 1/1000 second, you set the imaging devices to be sensitive to light for 1/1000 second. You have not changed the rate at which these exposure intervals occur–this is locked to the field rate of the television standard being used (60 exposures-per-second in North America, 50 exposures-per-second elsewhere).

In North America, the 1/1000 second exposure would mean that there is

$$1/60 - 1/1000 = 100/6000 - 6/6000 = 94/6000 \text{ second of time}$$

between the exposures. With rapidly moving objects in the scene, there would be some strobing from field-to-field. (The amount of strobing is dependent on the exposure time and the speed at which the object is moving through the scene.)

There are two basic types of shutters. For lack of more definitive terms, I'll call them physical and electronic shutters. A physical shutter is usually a rotating disc that can block the light path between the lens and the imaging devices. The rotating disc has a slice missing that allows the light to pass for a short period. An electronic shutter turns off the light sensitivity of the imaging devices.

On shuttered cameras that use tubes for imaging, there is a physical shutter. A physical shutter is similar to shutters used in motion picture cameras to shield film from light as the next frame of the film is positioned for exposure inside the camera. Because there is no easy way to turn off the light sensitivity of the target materials inside the tubes, the light must be physically blocked. Early CCD cameras also had the same requirement.

Newer solid-state imaging cameras now use chips that can be made insensitive to light except during intervals when the imaging is to occur. This technique can only be used with solid-state cameras where the sensitivity of the imaging elements can be electronically "biased" on and off.

Q: What do you look for when evaluating the picture created by a camera?

A: Regardless of the source of the picture (camera, VTR, character generator, etc.), the same basic procedure can be followed. View the picture from a distance of about two times the display height. (For example, for a display that is 20" *tall*, get 40" away.) If you are evaluating a camera, look at the same camera signal on at least two color picture monitors. If you are evaluating monitors, look at several video signal sources. Make sure that everything is clean, too. (There's nothing worse than making an evaluation with a monitor with a layer of dust or tar obscuring the picture.)

How does the picture "feel?" Does it seem sharp or dull? Do the colors look right? Is the picture stable and not moving under widely varying scene conditions? Remember, we are looking for first impressions. This is what the average of the great unwashed television viewing American public will think of the picture (before creating neurocide with a couple of beers).

Once you create that first impression, try to find out why it appears sharp, fuzzy, moving, etc. Sharp pictures mean that the transitions from bright picture areas to dark picture areas are clean and distinct. The bright should stop abruptly where the dark begins. The rate at which these transitions are created is associated with the limiting resolution of the system (frequently measured in TVL–TeleVision Lines–not 525 or 625 scan lines). The sharper the transitions, the higher the resolution and the cleaner the picture.

Fuzzy pictures can be caused by the inability to create sharp transitions or they can be caused by interference with the desired picture. Carefully look at the picture transitions for ghosts. Ghosts are an indication of termination problems, aperture or detail enhancement misadjustment, or problems in the design of the circuitry. Make sure that the problem is not induced in the cable or terminations being used for the evaluation.

Now look for any wavy lines dimly superimposed over the picture. This usually is the result of an interfering signal getting into the picture. This interfering signal can be induced by a stray radio waves (from AM, FM, TV, microwave, satellite, etc.), or they can be created with some types of processing. VTRs frequently induce substantial interference because of the way in which they must doctor the signal in order to record a portion of it!

Now carefully watch the picture as the scene changes. Look at the rate of decay of the bright picture area when going from a light scene to a dark scene. Look for any streaking (usually to the right) from a bright object in the scene. (Some cameras using "interline transfer" CCD chips may create a vertical streak from an unusually bright object. That's a problem with the design of the imaging chip in the camera.)

Now look at the color. This one can be a toughie because there are operator adjustments on cameras, recorders, TBCs, and color picture monitors that can fool you into thinking that there is a problem when there really isn't one. Make sure that the system (cameras and monitors are in black balance and white balance) before making an evaluation of color. I've been fooled many times–walk into your local TV store and look at the differences in color. Most of the differences are induced by different set-ups of the TV sets–not with some definite increase in color quality inherent in design. Remember, too, that ambient light under which you look at a picture monitor will change the perceived color.

Once you complete the up close and personal evaluation, look at the picture at a distance of eight times the screen height. This is the maximum viewing distance that is recommended by many "ex-spurts." Redefine your impression about the quality of the picture.

A couple of cautions here–camera original will always look best. Then comes first generation, then second generation, etc. Make sure that you are comparing apples-to-apples. Don't compare fourth generation S-VHS tapes with third generation ¾" or twentieth generation D-2. Determine how many generations that you need to do your job and then compare generation-to-same generation.

At the same time you confirm the number of generations, make sure that you are considering the capabilities of equipment currently in production, not some older equipment in which current improvements have not been installed. Don't let the wool be pulled over those peepers! (I am the resident cynic because I've had people try stunts like these!! *Caveat emptor!*)

Computers

Q: What is aliasing?

A: Aliasing is a distortion from a character or graphics generator that creates jagged diagonal lines in the picture. Alphanumeric characters also have many diagonal elements. Because of the nature of the beast called television, there will always be some aliasing.

Take a close look at the characters generated by your CG. Look particularly at the "C's" and "O's." Notice how jagged they are. The next time you see characters on your home television set, notice how jagged they are. The differences between the quality of the individual letters are indicators of the differences in price among the various units.

Anti-aliasing fonts are generated by high-end character generators to create very smooth, sharp letters. The CG equipment and the programmed instructions to the CG must be capable of working together to create these smooth characters. This all costs money. Again, you get what you pay for.

Q: What is a "paint" system?

A: A paint system differs from a character generator primarily in the number of "pixels" (individually discernable picture areas) and in the ability to address each of these pixels individually. In computer parlance, graphics employing this technique are said to have been generated by a "bit-mapped graphics" system.

Using a paint system, if you complete a circle with a wavy line, the location of each pixel that the line connects is remembered. If you can remember the location of each pixel of each line, you can establish a mathematical subroutine to determine the boundary of the circle. Using the mathematical model, any value of pixel that resides within the circle can be addressed. Now you can change the visible information within that circle to any desired color or luminance. This "fill" capability is the basis for many of the benefits of using a paint system.

You can also sometimes see aliasing problems in pictures from chip cameras. It usually shows up on nearly horizontal scene elements while the camera is panning across a scene.

Q: Why can't I use my personal computer as a character or graphics generator?

A: You can, but ...

As discussed in other installments of this continuing saga, the scanning rate of personal computers is not the same as the scanning rate of the television system. The scanning for television will not change–there's too many TV sets out there that have been designed to perform using the existing standards. The scanning for computers will not change–in the demented mind of computer designers, the higher the scanning rate, the higher the quality of their graphics.

Enter some of the white knights in shining armour with circuit boards that interface a computer with the television world. There's a lot of hubbub about "desktop video" and other such marketeer terms. Only a rare few of these interfaces really work. (Remember my earlier tirades about genlockers and the like?) Generally, by the time you buy one of these interfaces to magically transform your computer into a truly practical character or graphics generator, you're going to pay as much as for a low-end character or graphics generator.

I know, I know, you want to use the character generator functions sometimes and other applications programs (word processing, spreadsheet, etc.) at other times. Fine. Then you are probably going to wind up with a platypus that is either more difficult to operate (what function keys do what) or slower (the–computer–says–let–me–think–about–that–command–for–a–while–to–figure–out–what–you–really–want–me–to–display–on–the–screen) than a dedicated device.

There is one type of application for which these interfaces are truly appropriate. When you are trying to use statistics, computer-generated business graphics, or other microcomputer-based information in your television production, the interface approach really makes sense. In one system, I specified a plug-in interface board for a computer to air the database statistics of a basketball team. Here, an off-the-shelf computer economically manipulates the data providing an appropriate source of information that can be tapped for television distribution. (That same system *also* had dedicated character and graphics generators for graphics functions.)

Personally, I am convinced that you get what you pay for. For me, I use one machine for some computer applications and other machines for other computer applications. The computer between my ears has an easier time dealing with different keyboards than dynamically changing the keys on the same keyboard. This approach takes up more physical space but takes up fewer neurons in my only truly personal computer. (We dinosaurs are sloooow to change!)

Much ado about 'nutin, honey... [Thanks, Marvin!!]

Q: ***I just purchased a box of Kellogs Frosted Flakes and a box of Post Corn Flakes. What kind of cereal interface is needed? —Anthony "Tony" Tiger***

P.S. A prompt answer would be GRRRRRRREAT!!

A: Remember that a cereal interface needs to monitor its throughput to maintain a constant mass, not a constant volume. These things are, after all, sold by weight and not by volume and relying on bran names is not altogether appropriate.

To maintain a constant flow of mass, the cereal interface must properly address the needs of the source cereal as well as the destination cereal. The throughput efficiency of a cereal interface is often controlled by a "Mikey" protocol. If Mikey likes it, the mass will pass without further modification.

If you must pass the mass from one form to another, other significant modifications must be introduced. This becomes obvious when the Teenage Mutant Ninja Turtle standard must interface with the older Count Chocula standard. The Chocula standard has difficulty sucking the information out of non-human forms—Captain Crunch would serve as a much better source in this case. Such a difficult transfer, if not properly handled, will create a distortion bandwidth approximating the Frankenberry distribution.

The optimized transfer of cereal information must be accomplished bit-by-bit in bite-sized chunks. This is necessary to make sure that the 19 influences of the cereal interface provides the Special K-factor that provides a part of a balanced breakfast.

If too much information tries to go through the interface, a "puffed" condition will occur. In this case, the base cereal will take on a condition that will allow it to float longer in milk without losing its fibrous and crunchy structure. This is a particularly sticky problem when dealing with the "hot" cereal standards.

What we need is an *international* component cereal interface standard. Standards like Lucky Stars and Trix are already paving the way with their multi-colored component parts. Now we need to invent a process that will keep each component separated throughout the interface process–that's the only way we can keep it a regular world.

(Do you believe that I stayed up late at night to write this!!!)

Q: How can I get a cheap character generator?

A: The best way is to buy a used CG. Some of the older units don't offer the bells-and-whistles of the new units, but they often look better than some of the units using new low-cost approaches. Price will vary widely. Last year, I saw an old 3M D-2000 go for $80 at an auction. (I don't know if it worked, but I suspect that it did.)

The second approach is to buy one of the personal computers with a genlocker. You already know how I feel about this one. Price–$1500 and up.

The third approach is to build one. According to some of the books I have, you can "roll your own" for about $50, plus computer. You may want to look at the ***Cheap Video Cookbook*** or the ***TV Typewriter Cookbook,*** both by Don Lancaster and printed by Howard W. Sams. (Radio Shack used to carry some of these books, but I haven't seen them on the shelves in quite a while.)

Q: What does a scan converter do?

A: Scan converters are used to make an incompatible video signal compatible with television standards. Often, you will see scan converters used to take European-standard television video signals and turn them into North American standard television video. At other times, you will see scan converters connected to computers to take their weird video standards and turn them into something useful for television equipment.

Basically, a scan converter must be a very smart device. It must rapidly interpret incoming signal information, interpret it, then output another type of signal. It's kinda like the interpreters at the United Nations except the scan converter must eliminate or replace some of the signal information.

Let's take an example. Let's say that we want to take a

personal computer with EGA (Enhanced Color/Graphics Adapter) video output. The EGA outputs 21,850 Horizontal scans-per-second of video signal.

Television equipment uses exactly

$$525\ {}^{\text{horizontal scans}}/_{\text{frame}} \times 39.97\ {}^{\text{frames}}/_{\text{second}}$$

$$= 15{,}734.264\ {}^{\text{horizontal scans}}/_{\text{second}}$$

What do we do with the extra 6,116 horizontal scans that occur during every second of operation? Which of those horizontal scans will be deleted from the converted picture? Can some of the horizontal scans be averaged together to come up with a better finished picture? All of these questions must be answered.

The other problem with EGA scanning is that it is progressive, not interlaced. For every frame, the picture is scanned only once vertically. In television, each frame is scanned vertically twice (two fields-per-frame). Now we must be concerned about maintaining the interlaced relationships among the horizontal scans.

In general, the more intelligent the conversion process, the better the picture and the more expensive the scan converter. Bare bones conversion is provided by the under–$1,000 genlocker class. Averaging is provided by the $2,000–$10,000 class. Intelligent picture interpretation to improve apparent picture sharpness is the $10,000+ class of scan converter.

Another function of scan converters under some uses is to convert the signal from digital to analog. A CGA (Color Graphics Adapter) or EGA circuit outputs separate *digital* signals for the red, green, and blue video signals as well as an intensity signal. These signals must be converted into representative *analog* red, green, and blue signals for use with television equipment. Frequently, these separate red, green, and blue analog signals are encoded into one complete "encoded" analog video signal that contains most of the picture information contained in all the separate signals.

Q: How can you pull "composite sync" out of the 9-pin video out of a CGA card on a computer? The pin outs indicate "horizontal drive" and "vertical drive," as well as the R,G,B and ground signals, but no "composite sync". We have a couple pieces of equipment that will accept R, G, B, and Sync, including a Sony Mavica. --Louie Gorena of Pan American University

A: Television's "composite sync" is a combination of the timing information contained in "vertical drive" and "horizontal drive" (plus a few equalizing and serrating pulses thrown in for broadcast). Combining the two signals *could* be accomplished by using a simple combining network that would match impedance and provide isolation, *but* the scanning rate of CGA is not the same as television. Television equipment probably would become very confused about its scanning standards and cause the picture to roll, tear, and jitter.

The other problem with this approach is that CGA (computer Color Graphics Adapter) "R, G, B" signals are not the same as video "R, G, B" signals. The CGA output signals are of digital "TTL" (Transistor-Transistor Logic) configuration. Television equipment, unless specifically marked as having "TTL" inputs expect, *analog* R, G, and B signals.

In short, this one is going to take more than just a combining network; it's going to take one of those fangled computer "genlocker" circuits installed for the computer to talk television. (Remember that "video" is not necessarily "video!")

Q: FMI (for my information), what is VGA, EGA, CGA? — Wayne Byard, Video Dynamics, Inc., Miramar, Florida

A: VGA, EGA, and CGA are three of the more popular technical standards for computer displays. These standards represent the evolutionary standards that the IBM-compatible computer industry has adopted over the years. There are many, many more standards, current and obsolete, in the repertoire of the IBM-compatible computer industry (my computer display uses the IBM 8514 standard, yet another in the list of "standards"). Computers other than the IBM-compatible camp (Apple, Commodore, Atari, etc.) have adopted other standards.

As far as the IBM camp is concerned, in the beginning, there was monochrome (black-and-white). And the monochrome was with text. Then the monochrome begat the desire for cost-effective color graphics–the Color Graphics Adapter (CGA). Then there was a desire to improve, or "enhance" the quality of the color graphics–the Extended Graphics Adapter (EGA). Then IBM begat the latest "popular" standard with even more improvements–the Video Graphics Adapter (VGA). The VGA is the latest popular standard, although other higher-quality standards (8514, Super VGA, Enhanced VGA, E.I.E.I.O.) are being bantered about in the computer industry.

All of the computer standards are derived from combinations of separate Red (R), Green (G), Blue (B), and Intensity (I) signals. Some of the differences among the popular standards are as follows:

Standard	Signal Types	Vert Rate	Horiz Rate
CGA	R,G,B, and I Digital	59.92	15700
EGA	R,G,B, and I Digital	59.7	21850
VGA	R,G,B, and I Analog	Several	Several
NTSC (TV)	Encoded Analog	59.94	15734

None of these computer video standards are directly compatible with the television video standards–you cannot simply connect a computer to a television display. To use material created within these computer standards in your television program requires at least a "genlocker" box and at best a sophisticated "standards converter." There is a wide array of "genlocker" plug-in cards and boxes that purportedly will interface computer video standards with television video standards. You get what you pay for. I've seen all kinds of problems with these types of boards–from *gross* timing errors to flicker and strobe in the finished picture. To the best of my knowledge, there simply isn't a cheap way to do it–yet.

Looking in my crystal ball, I think we will have a common computer/television standard emerge within the decade. This standard will bypass the HDTV (High Definition TeleVision) analog standard that is now languishing in various politically based committees. Only then will we have a true "desktop video" environment. (Even though there may be the environment in which to work, hardware does not a professional communicator make! As with so much of this technology stuff, the capability of the hardware is outdistancing the ability of many of the hardware purchasers to communicate.)

Q: What is the best way to shoot a scene with computer monitors in the picture?

A: The *easiest* way to shoot a scene with computer monitors in the scene is to eliminate the possibility of seeing the computer display. The difference between the vertical scanning rates of most computers and the vertical scanning rate of television will usually create dark horizontal bands moving vertically through the picture. The rate at which the dark bands appear and move is dependent on the difference in frequency.

The best (and most complex and expensive) method is to change the vertical scanning rate of the computer displays to match the scanning rate of television. It may take a scan converter locked to the master sync generator being used by the television equipment. Remember, another scan converter will be required for each computer screen.

As an alternative, try to arrange the scene with only static characters and images expected to be displayed on the computer screen. Then you can overlay a plastic sheet with the desired display on monitors that have been turned off. No vertical scanning of the computer monitor and voila, the dark horizontal bars in the finished scene disappear.

Another alternative is to use chroma key techniques or digital video effects (DVE) to insert the desired display into the monitor. The DVE source must have a vertical scanning frequency that is compatible with television vertical scanning. Perhaps a little bit of animation from a character generator or a few minutes of recording from a computer that *is* television compatible would suffice.

In general, you can't count on computer picture monitors to be compatible with the television system. The best policy is to simply keep them out of the picture. This is particularly true when the computer displays are used as terminals for access to a large "mainframe" computer.

Q: Shooting with a broadcast-quality camera, how does one eliminate the "rolling bar" when shooting picture monitors, CRTs and color graphic display terminals?

A: The "rolling bar" is a dark horizontal bar (through which you can still see the normal picture) that moves vertically through the picture. This bar appears when the scanning of the camera is not synchronous with the scanning of the display. The rate of motion of the bar is determined by the amount of difference between the two scanning rates.

The bar may be eliminated by "simply" making the camera and display synchronous. This can be *tough*. Television equipment is designed to meet the scanning specifications specified by the Federal Communications Commission. Computer equipment isn't. Computers don't meet any universal standard. If you're shooting television video, you have no choice of scanning standard; therefore, you must change the display.

Computer displays usually have vertical scanning rates from 49.8 fields-per-second to 60 fields-per-second. This vertical scanning rate must be changed to 59.94 fields-per-second to be compatible with broadcast color video.

Computer displays have horizontal scanning rates from 15,700 scans-per-second to 31,500 scans-per-second and above.

This horizontal scanning rate must be changed to 15,734.264 scans-per-second to be compatible with broadcast color video.

How do you change the scanning rate? You use what is essentially a "frame store" and read out the information that you need to create a usable picture. This particular kind of frame store is not what you find on the shelves of your local video store. This special frame store must contain a program to decide which picture details are to be kept, which details need to be thrown away, and which details need to be repeated to fill in lost picture areas. For universal "converters" from odd-ball computer scan rates to sane, stable television scan rates (can you tell where I'm coming from??) you can expect to pay anywhere from about $7,000 up.

This is the same signal processing philosophy that is necessary when converting an NTSC-M (525/60) video signal into a PAL-B (625/50) video signal, or vice versa. We television folks have called this "standards conversion" (of scanning) and "transcoding" (of color) for years.

One other challenge that must be met by the frame store is the aspect ratio of the display for which the computer "video" signal is designed. Television displays use an aspect ratio of 4:3 (four units wide by three units high). Computer displays usually have some other ratio, sometimes even 1:1. The frame store decisions of which picture elements to keep, use and repeat become even more important in this situation.

The computer folks are now coining the term "genlocker." This is a bastardized video term that some manufacturers are using to describe a piece of equipment that takes the "video" signal output from a computer and converts it to a "video" signal usable by television equipment. So far, the genlockers I've seen are in the under-$1000 class and are dismal failures in fulfilling their promise. Apparently, many of the manufacturers of these products ignore the scanning, delay and bandwidth specifications that they profess to meet. Computer-specific genlockers in the $1500-up class are getting mixed reviews from users.

Once the computer signal has been converted to a television

broadcast video signal, it may be directly recorded like any other video signal. If you're trying to compose a shot of real objects (including talent) with a computer display, you can then synchronize your camera (by genlocking to the converted computer signal) with the display and completely eliminate the "rolling bar."

Now for a couple of lower-cost options. These may or may not work. The first involves changing out the display's display. By installing a CRT with phosphors that have a longer brightness decay time, you may be able to mask some of the effects of the non-synchronous operation.

The second approach requires readjustment of the camera. By increasing the lag of the tubes in the camera, you may also be able to reduce the "rolling bar" effect. With a vidicon-equipped camera, it's easy to adjust the TARGET to get this effect. With other tubes, like the Plumbicon®, Saticon®, or Newvicon® you've got to be a lot more careful. I suggest that your technician with tool kit be available if you're going to readjust anything other than a vidicon camera.

Neither of these two approaches will eliminate the "rolling bar", but will simply reduce its effect.

Q: What is meant by 24-bit, 32-bit, 64-bit, etc.?

All this nonsense refers to the number of individual bits of information that can be simultaneously processed in a piece of equipment. It's analogous to the number of parallel lanes that a highway has in each direction. (I wish I could find some 24-lane highways here in the D.C. area!!!) It's also analogous to stereo being a two-channel system with two lanes of highway capability.

The greater the number of channels, the faster the digital information can be conveyed. In a 64-bit system, 64 pieces of information are gulped down in one swallow.

It's like visiting an Irish pub. We can quaff down a pint of ale (all the available information) in big gulps (64 bits at a time) until we have an empty mug. We can sip the ale in smaller gulps (24 bits at a time) until we have an empty mug.

If, in both cases, we use the same clock to determine the rate at which we gulp, the big gulper will finish the brew much sooner than the little gulper. In this example, we are determining the rate at which the ale is transferred for processing. Remember that something must be done with the ale once processing has been finished.

When information can be processed in a speed manner, we begin to enhance the sensitivity of the hardware to changes in the information. We can process more minute information in a timely manner. Let's assume that we are dealing with digitized picture information. To keep up with television, we need 29.94 complete pictures processed each second. Let's assume we're using all of the 525 horizontal scan paths to convey actual picture information. (Normally about 40 of the 525 horizontal scan paths are "blanked" to allow time for vertical retrace.) Let's also assume that we have 1024 individual picture elements on each of the horizontal scans. This means that, in order to keep up with the television standard, we need to process

$$29.94\ \frac{\text{frames}}{\text{second}} \times 525\ \frac{\text{horizontal scans}}{\text{frame}} \times 1024\ \frac{\text{pixels}}{\text{horizontal scan}}$$

$$= 16{,}095{,}744\ \frac{\text{pixels}}{\text{second}}$$

Over 16 *million* pixels-per-second. Whew! This means that, with a clock that produces 16 million ticks-per-second, we can process one pixel per tick (or tock). This also means that the number of available paths through which the signal is processed limits the quality of the picture.

Let's assume that you have a system with 16 parallel paths for communication. For our example system to keep up with the rate at which pixels appear, an individual pixel could have only 16 digital bits of information. This means that only

$$2^{16} = 256$$

individual shades of individual colors that can be discerned in a picture. Increasing the number of parallel paths will produce the following results

$$2^{24} = 16{,}777{,}216$$
$$2^{32} = 4{,}294{,}967{,}296$$
$$2^{64} = 18{,}446{,}744{,}070{,}000{,}000{,}000$$

The greater the number of individual colors, the greater the perceived picture detail and the more lifelike the picture will appear.

There are some systems that use 16-bit architecture within the computer, but allow the use of a 24-bit (or higher) plug-in card. In these cases, the 16-bit machine provides some of the control and some of the communications capability, but a large portion of the image manipulation actually occurs on the card itself. The card has its own "image processor" separate and apart from the "central processing unit" (CPU) of the computer. The image processor may use a clock that is slaved to but faster than the CPU clock. It's like using the commuter lanes in an urban highway to create a better image.

The next generation of computers will use "parallel processing." Instead of designing one large highway, there will be several individual highways, each with its own signal processing and manipulation. A 64-bit parallel processing system will have 64 CPUs operating with reference to the same clock. Each CPU will concentrate on maximizing the time used to think about and manipulate only one channel of information. When we get to that point, we'll really begin to see some decent graphics and imaging systems. Until then, we're saddled with the slow single processor technique.

Parallel processing systems are now available on the market, although they are still rather rare. Given 10 years, in our home TV sets, we'll be able to perform what are now considered sophisticated image manipulation, stabilization, and processing. Then it'll make a 24-bit Amiga (or IBM, or Apple, or whatever) look like the original Apple looks today. Nothing lasts forever – such is the price of progress.

Q: What is meant by "24-bit" processing in computer imaging equipment?

A: This is a clarification of the information in the July, 1991 article. (I know, I know. It's about time!) The typesetter missed the exponents on the terrible twos in that article.

Remember that the terminology refers to the maximum number of bits (time slots in a digital signal) that the equipment can process for each individual picture element ("pixel"). For most equipment, you can convert from the "x-bit processing" hype in manufacturers' literature into a number of colors by using the following formula:

$2^{\text{number of bits}}$ = number of available colors in the palette

For instance, in a "24-bit" system, there are

$$2^{24} = 2\times2$$
$$= 16{,}777{,}216 \text{ colors available in the palate}$$

(Notice that there are twenty-four "2s" in the formula.)

To save you the math (I know how much you just *love* math), here are the numbers:

2^8	= 256	2^{17}	= 131,072	2^{26}	= 67,108,864
2^9	= 512	2^{18}	= 262,144	2^{27}	= 134,217,728
2^{10}	= 1,024	2^{19}	= 524,288	2^{28}	= 268,435,456
2^{11}	= 2,048	2^{20}	= 1,048,574	2^{29}	= 536,870,912
2^{12}	= 4,096	2^{21}	= 2,097,152	2^{30}	= 1,073,741,824
2^{13}	= 8,192	2^{22}	= 4,194,304	2^{31}	= 2,147,483,648
2^{14}	= 16,384	2^{23}	= 8,388,608	2^{32}	= 4,294,967,296
2^{15}	= 32,768	2^{24}	= 16,779,216	2^{33}	= 8,589,934,592
2^{16}	= 65,536	2^{25}	= 33,554,432	2^{34}	= 17,179,869,184

Q: What kind of personal computer makes the best graphics generator for television?

A: All of the above. There is nothing really sacred about IBM, or Apple, or Amiga that makes one significantly better for creating graphics. It's mostly which platform has the supporting hardware that fulfills your graphics application. So far, to me it appears that the Amiga got a head start in pushing the television graphics idea a little more strongly than the other manufacturers, but it's primarily just a matter of time before the competition catches up. All the manufacturers see our market as having enough potential growth to try to develop product to serve our needs.

As you look at the products, remember to consider them in context with a television system. In addition to a particular plug-in card, you may need other plug-in cards or other equipment just to make it work.

One of the most confusing aspects of reading literature about computer uses in television has to do with how the final product is presented. There are lots of products that simply make your computer screen look good. There are lots of other products that simply make your printed output look good. There are lots of other products that can interface your computer with your television system.

Unfortunately, some of the computer/television ads don't tell you what a product is intended to do. If it says multimedia, I tend to believe that the product makes the computer display look good. Frequently, you can't record the resulting signal with a conventional video tape recorder. You can't look at the signal on a color television picture monitor. You can't integrate that signal into your television programs. You can store the signal in conventional computer memory. You can use a non-standard display or data projector that will present your work.

You cannot assume that "video" means television. Computer companies are using the term to mean any scanning, signal domain, and transfer method (including the traditional television standard) that is convenient to their proprietary scheme of things. The term video has now been weakened from the single definition that we enjoyed for years before the computer types discovered it.

Even the specifically-defined "RS-170A" doesn't mean anything to many of the computer product manufacturers. I've talked with several computer manufacturers and they simply cannot understand the underlying technical concepts of RS-170A. (Such things as the subcarrier-to-horizontal phase relationships in RS-170A are ignored in the testing of some of the "RS-170A" products.) In my opinion, if a manufacturer claims compliance with RS-170A and if the product doesn't fulfill *all* of the requirements of RS-170A, it is fraud. This question becomes critical if you try to use the product in high-dollar facilities using professional video equipment that is sensitive to SC/H phasing (like BetaCam®, 1"-C, D-1, and D-2). If you buy a product with a smaller system now, you want it to be capable of supporting a larger system later.

Q: How can I keep track of all the new "multimedia" products that are appearing on the market?

A: I dunno. It's as confusing for me as it is for you.

I try to categorize the products. Since I'm primarily involved in TV, I first categorize the products by the way in which they process TV signals. Can the video signal be composite (AKA NTSC, RS-170, RS-170A, encoded), Y/C, R/G/B or Y/R-Y/B-Y?

The second step is to figure out which computer platform(s) the product supports. This could be Amiga 2000, Apple Macintosh, Apple IIe, IBM-AT, IBM-MCA, Univac, or a sacroiliac (a backbone system). Does it require a large amount of memory or a huge hard drive?

The third step is to determine how the product interfaces the television system with the computer. Is the interface for control only (like an edit controller)? Is the interface simply a switcher or mixer that is controlled by a computer? Does the interface take a television signal and convert it into a computer signal (like a frame grabber)? Does the interface provide a standard television output from the computer? The question of whether the product accepts a television signal as input or if it provides a television signal as output is critically important–there are many products that provides only an input to OR an output from the computer.

The fourth step is to determine what, if any, signal processing is performed by the product. Is time base correction, freeze frame, character generation, visual effects, aural effects, etc. performed by the product? With what quality are these operations performed? How fast does it perform these operations? How easy is the product to use? (Despite the advertising, there are a lot of products that do not create picture or sound that are appropriate for serious production! You *must* take a product for a test drive under the conditions in which you plan to use it!!!!!)

The fifth step is to determine the physical attributes of the product. Is it a plug-in board, an external box, or both? What kinds of connectors are on it? What environmental concerns (airflow, humidity, etc.) are voiced by the manufacturer?

Last, but not least, ask "How much is that doggey that makes the window?" Ugh!

Well, here I sit at 32,000 feet on a Sunday morning pounding away on my laptop. I just enjoyed a magnificent butterscotch sundae and watched the "Minding Your Business" program on the miniature silver screen. In that production, I saw something that sent my adrenalin level soaring–a software product review. This product was touted to be the ultimate product for the conscientious writer. Indeed, it was implied that no good writer could conceive of any gramatically correct verbage without it. It so happens that I've used this particular product for several years and have repeatedly fallen for the marketing nonsense. The last time that I used the product (two months ago), it corrupted my file so badly that the final report was rejected by a client. Moral to the story–don't fall for marketing claims without thoroughly checking them out. Chalk one up for my cynicism!

This leads me to some universal claims that are constantly bantered about the industry, but that turn out to be true less than 50% of the time:

1. "It is software compatible" ... (as long as . . .and you don't have garlic for lunch).
2. "It is hardware compatible" ... (as long as you use a model ... on alternate Valentine's Day).
3. "It meets RS-170A specifications" ... (if you look only at a particular portion of the specifications and ignore the obvious fact that it doesn't work for your application).
4. "Resolution greater than competing products" ... (if resolution is considered to be a function of temperature).

The facetious comments in parentheses are mine, but they represent the way I feel at trade shows. *Nothing* from *anybody* is to be believed at face value when talking about implementing a particular piece of equipment in your particular operating environment!!! It's not that the manufacturers are always out to deceive you–it's that the salesman's job is to sell you a product, not to help you communicate via television. In the last couple of months of ***Byte*** magazine, there has been an outpouring of the frustration of interfacing computer hardware pieces and software into a system. Does this sound familiar? We've been echoing those same sentiments for years in the television industry!

Basically the ***Byte*** authors feel that software manufacturers are abrogating their responsibility to the industry by shipping incomplete, nonworking, difficult-to-use products for the sake of profit. Many manufacturers in television, as well as in computers, could not care less about the end-users of their products. To them, the end-users of video equipment make up a market of individuals that are largely naieve in business and technical values. How many times have you called a manufacturer, to be told that "we never tried it that way before" or that "it can't be done with your equipment," or that "you have to sit on a right-handed blue toilet seat" for the system to work properly.

I believe that the only reason that we put up with this abuse is because that we (including me), as an industry, refuse to consider the complex technical issues in our industry. We're intimidated about all this technical gobbledygook. It's time for us to start turning things around. But how?

Television is a multimillion dollar-a-year business, but it is very fragmented. There are thousands of "one-man shops" that have each spent $100,000 or more. How many of those shops have equipment that "never worked right" or that failed prematurely. What did they do? Usually they shoved the equipment into a closet and wrote it off to experience.

In my short lifetime, I've experienced

- graphics generators that promise 256 colors but can only display 16 colors in NTSC;
- video tape recorders that constantly and reliably damage tape;
- tube cameras that alter registration when you press on the sides of the case; and,
- equipment that requires routine replacement of parts costing almost as much as the entire original unit.

In *each* of these cases, the manufacturer shrugged his shoulders and walked off, leaving me or my client holding the bag. Whatever happened to the old notion of owning up to one's responsibility? I'm not purrrfect, but I stick with a problem until I am satisfied that it's substantially correct or I am approaching my limits of sanity, even if it means losing money. I owe that to the industry.

How many of these "free" magazines do you get? Who do you think pays for them?? Count the number of pages of advertising versus the number of pages with editorial comment? Now who do you think pays for them?? Consider that there are considerable editorial constraints imposed and demanded by the advertisers. Until you, as a user, are willing to pay for the dissemination of information, this is the way it will be.

All that advertising does one thing–it creates the impression that there is a huge market for a product even if the demand is limited. If you don't buy a particular type or make or model of equipment you are a nerd, dweeb, or worse. Forget fiscal responsibility. Forget the fact that we're in the business to communicate. Forget the fact that the equipment that we purchased last year is just now beginning to work as a system. Forget the fact that we may already be communicating to the utmost efficiency. *BUY. BUY. BUY. BUY.*

What are some of these markets that looking for a place to happen? In my opinion (and the basis for my opinion), they include:

- Electronic Still Production (suited for applications where a mediocre-quality still image needs to be instantly transmitted);
- Cheap Computer Frame Grabbers (where a low-quality TV picture is imported into a higher-quality computer display);
- TV-in-a-computer adapters (why bother???); and,
- Interactive Video (spend ten times as much money as what would be required to handsomely pay a gaggle of good teachers or to man an array of information booths).

Consider me a heretic, if you wish, but I am an opinionated b*@#&^^!!! Let's own up.

Users should own up to their lack of knowledge so that they can ask those all-important "dumb" questions and know when they get equally "dumb" answers. We're scared of equipment. We're repeatedly told that we'll "break it" if we try something innovative. We're naive about what it really takes to communicate via the video signal–the most complex signal in all of electronic communications.

Manufacturers should own up to their motivation to sell products at the expense of developing a market that can find innovative applications to the equipment being offered. Cut the B.S. Couch product features in terms that are universal in the industry. "Low capacitance bias time-division multiplexing of the CMOS device in a ECL monostable multibrator to accomplish seamless inter-scene transitions without the need for a D9" doesn't mean a thing to me. The ultimate question that should be answered is "what will it buy me???"

Q: What is the difference between a "frame grabber" and a "digitizer?"

A: Well, they're kissin' cousins. They both accept a television-type video signal and convert it into a digital computer-type signal. Once in the computer, the signal can be manipulated to create special effects, airbrushing, etc.

The generally accepted difference (there is nothing that is certain in the computer definition business) is that a "frame grabber" can capture a video frame from a moving image and a "digitizer" requires a still image. It all boils down to how fast that the card can convert a television frame into computerese. A frame grabber can do it in about second, but a digitizer takes longer.

Q: A company by the name of New Media Graphics talks about encoding video into a YUV file format. What do they mean? –Roy Moglia, St. Petersburg, FL

A: I wasn't quite sure, so I telephoned Jim Beveridge at New Media Graphics (508-663-0666) to get the straight poop. The Y,U and V are three video signal components used in PAL (European) encoding. "Y" means "luminance" (exactly the same as "Y" in NTSC terminology). "U" and "V" are two components of chrominance (color information). Mathematically,

U = Blue Video Signal - Y
V = Red Video Signal - Y

(I've always said that PAL was a kissin' cousin to NTSC.)

For each pixel, New Media Graphics stores some bits of information containing the Y value, other bits of information containing the U value, and still other bits of information containing the V value. As long as the scan rate can be adapted, it's easy to convert from Y/U/V into NTSC.

Scanning rate is another matter. Jim indicated that the conversion from 625 scan-per-frame (European standard) video into 525 scan-per-frame (North American standard) video is accomplished by selectively dropping 100 of the scans (chosen across the entire frame) in each frame of video.

Recorders

Q: Why does my normally-healthy recorder act up when I have it on a service bench?

A: The light from the overhead shop light is probably fooling the end-of-tape sensors into thinking that the end of the tape is at hand. These sensors often use visible light to sense the presence of the clear plastic head or tail that is spliced to the tape in the cassette.

Q: What is the difference between "Y/C" and "Y/688?"

A: "Y/C" refers to black-and-white luminance (abbreviated "Y") and color information called chrominance (abbreviated "C"). Chrominance has a standard frequency of 3,579,545.4545 cycles-per-second (3.58 MHz) in North America. There are two cables involved, one for Y and one for C, instead of the one cable for both Y AND C in NTSC.

The "Y" in "Y/688" refers to the same luminance as described above. The "688" refers to basically the same chrominance color picture information, EXCEPT that the frequency is 688,000 cycles-per-second. This "dub" standard is found in ¾" and BetaMax® VTRs. (One cable sheath through which the two cables pass is frequently found in Y/688 systems.) Similar standards, but with different frequencies, are found in VHS and 8mm machines. (I haven't seen these signals on output connectors—yet.)

Theoretically, you could heterodyne the Y/C signal into Y/688 (and vice versa) very easily. This allows maximum-quality small interformat editing (provided the edit control functions are adaptible.)

Q: I have considered editing up from VHS to ¾" as an edited master, then dubbing back down to VHS for my copies. Will this give better results than going VHS for the entire edit and dub? —Larry Brandt, Irvine, California

A: This one's a toughie. It's a subjective call. For backup, I contacted Sol Benatar at U-Edit Video in beautiful downtown Richardson, Texas.

You can certainly try your procedure, but I doubt that you'll see too much improvement. The limiting factor is the VHS format. The distortion artifacts in this format are worse than the distortion artifacts in ¾". Because the VHS is the weak link and still plays an important part in your procedure, Sol and I guesstimate about a 5%–10% improvement in subjective picture quality would be the maximum you could achieve over a VHS-VHS-VHS post-production procedure.

What would really be interesting would be to play your VHS production master on a Super VHS source deck (with Y/C outputs) and convert those signals to Y/688 for use with the ¾" "dub" input. Then pull the VHS distribution copies from the ¾" edit master. I would guesstimate about a 20–30% improvement in subjective picture quality over VHS-VHS-VHS.

Q: What is an "address track?"

A: An address track is an area of video tape that has been set aside for longitudinal time code (LTC). Normally not suitable for recording any other type of signal, the address track sometimes shares its space with other signals.

On the professional ¾" format, the address track is placed in the same area as the vertical sync pulses of the video signal. The time code signal is recorded "deeper" into the tape and at a different magnetic azimuth to the video signal.

Other professional formats, like BetaCam®, M-II, 1"-C, *et.al.* have areas of the tape specifically set aside for time code use.

Another type of time code, Vertical Interval Time Code ("VITC"–pronounced vit-c) is added to the video signal before both are recorded in the video tracks of the tape. VITC is not directly compatible with LTC, but interpreters are available to convert between VITC and LTC.

Q: After much pain and irritation, we have learned that audio channel time code and address track time code differ in "shape." What is meant by "shape?" —Chris Mosio, Goal Productions, Inc., Altadena, California

A: This was a new one for me. I called Andrew Simon and Gerry Lester at Adams-Smith, Scott Gray of Gray Engineering Laboratories, Inc., and Kevin Walder and Crit Taylor of Fast Forward Video for an explanation. Basically, the "shape" of the time code refers to the "rise time" of the time code signal. This rise time determines how fast the digital time code signal changes from "low" -to- "high" voltage states (and vice versa).

According to the SMPTE specifications, time code does not have an exactly rectangular waveform. As shown in Figure 6, the time spent in transition between the 10% voltage and 90% voltage amplitude must be between 20 and 30 microseconds (0.000020–0.000030 seconds). If the rise time is too short, the odd harmonics of the signal increase to the point of covering too broad a spectrum and can create interference with other video, audio, and data signals. If the rise time is too long, the time that the signal transitions from low-to-high cannot be accurately determined.

What causes the rise time to change? Usually it's a matter of matching the impedance and balance configuration between the time code generator and the VTR or between the VTR and the time code reader (that may reside in the editor). If you're having time code problems and don't want to learn about such stuff (or research my previously published excellent articles—hiss, boo, applause) simply try reversing the conductors in your connections. Keep in mind that address track circuits may or may not have the same impedance and balance configuration as audio track circuits on the same VTR!

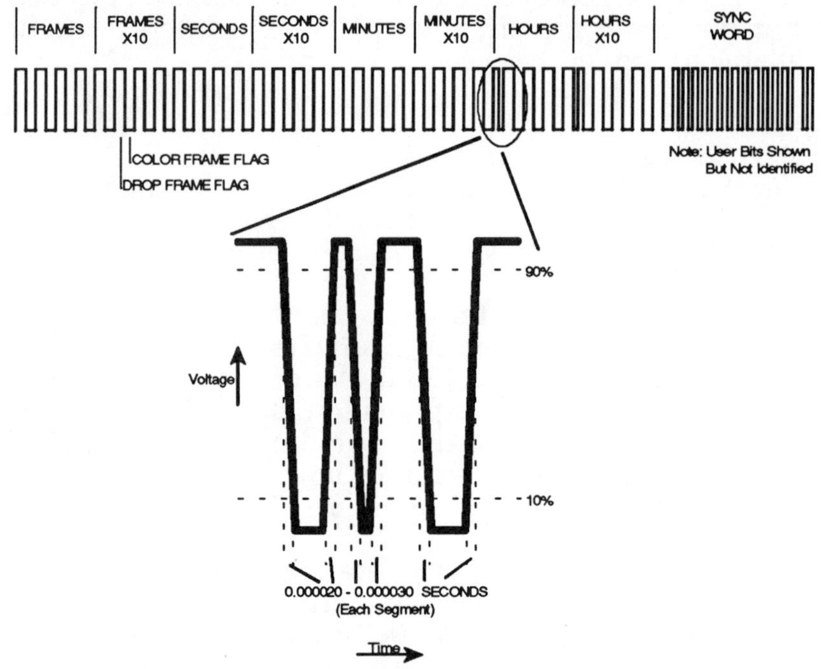

Figure 6
SMPTE Time Code Rise Times
(Longitudinal Time Code Shown)

If reversing the conductors doesn't make the system work, look at the specifications for the output of the time code generator and the input of the VTR (and between the output of the VTR and the input to your time code reader). If both say balanced you're OK. If both say unbalanced you're OK. If one says balanced and the other unbalanced, install a "balun" transformer in the line.

Some older VTRs require you to install wide-band time code address track circuits. Other VTRs require a time code level reading of +4 on the VU meter during recording. Play around a little; BUT, pay attention to your video signal, picture, audio signal, and sound to see if you're messing anything up.

Some manufacturers offer an internal adjustment between a time code shape generated to specifications and a sharper shape (about five microseconds rise time) for use when the specifications are not tight enough for accurate time code detection in a given situation. This is not a normal operating mode; but, if it works, fly with it.

Q: What is "color under?"

A: Color under is a term that is virtually synonymous with "heterodyne" in the video tape world. In order to get the color information (chroma) through the recording process, the frequency is changed from the normal 3.58 MHz into another, lower frequency. VHS, Super VHS, Betamax®, ED Beta®, ¾", and 8mm are the currently popular formats using the "heterodyne" or "color under" technique.

Within the VTR, standard NTSC chroma information (all of which has a frequency of 3,579,545 cycles-per-second) is converted ("heterodyned") within the recorder. In VHS and Super VHS, the new chrominance frequency is 629kHz. In Betamax®, ED Beta®, and ¾" the chrominance frequency becomes 688kHz. (Does "688" sound familiar? That's what they're talking about in the "Y/688" dub mode. The "Y" stands for luminance.) In 8mm, the heterodyned chrominance frequency is 743kHz.

Not only is the chrominance portion of the video signal manipulated, but the luminance is also processed before it gets recorded on the tape. A graph of the signal as recorded on a ¾" tape is shown in Figure 7. Notice in the diagram that the horizontal axis is frequency and the vertical axis is signal power. The luminance is frequency modulated to a range of frequencies that is higher than the heterodyned chrominance information.

The term "color under" refers to the relationship of the heterodyned chrominance frequency relative to the frequencies at which the frequency modulated luminance is recorded on the tape.

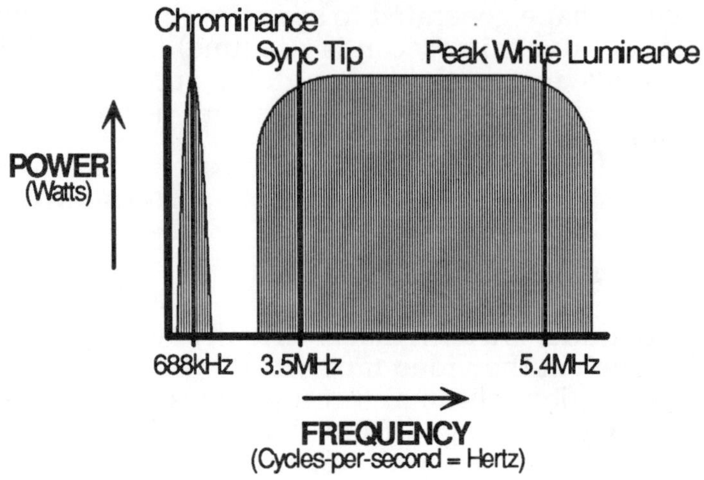

Figure 7
Graph of Signals Recorded in the Video Tracks of ¾" Tape

***Q:** Is it true that the narrower the width of the tape the more time base correction that will be needed? —Chris Pillar, Abbott Loop Christian Center, Anchorage, Alaska*

A: This question contains quite a wallop! In general, it is true that the narrower the width of the tape, the more time base error is induced; however, there are many more factors involved behind the scenes. Time base error happens whenever the speed of the video signal as played back does not equal the speed of the video signal when it was recorded. This means that the speed of the video head past the tape must be *exactly* the same during playback as it was during record–even if the tape is being played back on a VTR that is different than the record VTR.

The video signal is recorded in video tracks (paths of the video heads as they speed past the tape). Because of the way that the tape winds around the transport and the video heads, the tracks appear at a slight diagonal to the edges of the tape. In general, the wider the tape, the longer the video tracks.

The longer the video tracks, the smaller the effect of minor speed variations, and the less time base error that is created. So, on the surface, your assumption about the correlation between tape width and time base error *appears* to be correct. Now let's introduce some of the other factors that affect time base error in a given video tape format and I think you will begin to see the complexity of the subject.

First, let's look a "gyroscopic precession." The video heads on all modern video tape formats are mounted on a spinning cylinder. The spinning cylinder acts just like a gyroscope. (Remember high school physics–if you don't, go to a local toy store and buy a toy gyroscope and play with it.) Basically, any time that the gyroscope is forced to move in space, it resists the applied force. When it resists the force, it induces a slight change in rotational speed. In a VTR, a change in speed of the video head cylinder creates time base error. The external forces are induced whenever the VTR is moved. Imagine the speed variations that occur when you're banging a portable VTR or "Camcorder" around!

Another good source of information about gyroscopes are books about instrument flying–there's usually a dandy explanation in the section about the artificial horizon, flight directors, and inertial guidance systems. (Isn't it amazing where I can dig up stuff?) This is a good enough reason to go take some flying lessons–at least a flying leap!

The amount of speed variation depends on the mass of the video head cylinder. The larger and "heavier" the cylinder, the greater the amount of speed variation that is induced from a given force. A Super-VHS cylinder is bigger than an 8mm cylinder, sooo . . . As an experiment, place a VTR on a table. Enter the play mode and watch the playback picture. Now pick up the VTR by the sides and rotate it about thirty degrees in the horizontal axis (so that the bottom of the VTR remains parallel to the table). Watch what happens to the picture as you move the VTR. It'll move. Depending on the picture monitor, the mass of the spinning video head drum assembly, and the speed of the VTR rotation, the picture may move a little or it may move a lot. This picture motion is the picture monitor's way of showing you time base error.

There are other gyroscopes inside the VTR. The tape on the supply reel, the tape on the take-up reel, the supply hub (that drives the supply reel), the take-up hub, tape guides, etc. Although their effect is not as pronounced as with the video head drum assembly, they do induce some time base error.

Now look at the mass of the tape itself. It takes more force to move a greater mass. In terms of raw physical theory (Force = Mass × Acceleration), tape that is ½" wide will have a higher mass than tape that is 8mm wide. This means that the resulting effect on the time base from a particular magnitude of force ("Acceleration" in the formula) will be less on a smaller tape.

Although it may be intuitive that wider tape widths reproduce a more stable video signal time base, it is not entirely true. Both 8mm and ½" wide tape can make it in the industrial world, if time base is the only factor. To the best of my knowledge, none of the basic VTRs incorporate any kind of special "jitter control" to reduce the magnitude of time base error.

Keep in mind that time base error is a factor primarily in the stability of the position of a played back picture (and editing the played back picture). It has little, if anything to do with the overall picture quality of a played back picture. There are other factors at work to determine the overall picture quality.

Don't let a single, particular parameter or specification (especially time base error) be the determining factor in the selection of a tape format. (There's too much *dis*information flying about to make any single source infallible.) Use some demonstration equipment under the conditions that you expect to later use the purchased equipment. Make sure that you are comparing the latest versions of the equipment–improvements are made every day and direct comparisons will be invalid if you don't compare what you can actually have delivered today. If it fills the bill–buy it. If it doesn't perform to the level of satisfaction that you require and expect–look on.

Q: What video tape formats are popular and available today?

A: I like to divide this down into groups of formats that use similar video signal processing in the recording process. Here goes:

<u>HETERODYNE CHROMINANCE VIDEO PROCESSING</u>

8mm, VHS, VHS Hi-Fi, Hi8®, Super-VHS, ¾", ¾-SP®

All these formats use the same *basic* method of recording the color portion of the signal onto the tape. (BetaMax® and its derivatives fall in this category, but . . .)

<u>COMPONENT ANALOG VIDEO PROCESSING</u>

BetaCam-SP®, M-II

These formats record three parts (Luminance, Red minus Luminance, and Blue minus Luminance) of the video signal in three different areas of the tape. These formats were initially designed for high-quality field acquisition.

<u>DIRECT SIGNAL VIDEO PROCESSING</u>

1"-C

This format records a version of the *entire* video signal without separated chrominance processing.

<u>COMPONENT DIGITAL VIDEO RECORDING</u>

D-1

The D-1 format is *expensive* for the time being, but it's still the highest possible quality available from today's commercially available recording technology.

<u>COMPOSITE DIGITAL RECORDING</u>

D-2, ½" Digital

The D-2 format appears to be positioned to replace 1"-C in a few years.

***Q:** What are the specific technical advantages of the* **composite** *output of M-II and BetaCam-SP® over U-Matic-SP® in an A/B edit system? Does an interformat system (U-Matic-SP®-to-M-II or BetaCam®) have any advantages over a straight U-Matic-SP® system? —Scott Frederick, Baylor College of Dentistry, Dallas, Texas*

A: Before I start weaving an answer to this mammoth question, let's make note that the composite output from *any* VTR is not ever going to be as good as any available component outputs from the same VTR. The currently allowed encoding scheme for NTSC creates so much distortion (compared to the standards allowed by technology that has recently become available) that almost anything is an improvement.

RULE #1
A COMPOSITE OUTPUT IS GOING TO PRODUCE LOWER PICTURE QUALITY THAN *COMPARABLE* COMPONENT OUTPUTS.

The composite output a VTR designed for component operation was installed primarily for monitoring of the output signal on a readily available "inexpensive" composite picture monitor. (If the composite output were not there, you would need a monitor with component inputs just to look at the picture. This is not out of the question but component input monitors are not as readily available as composite input monitors.) I have heard of someone wiring their entire BetaCam® system for multiple generation capability using only composite inputs and outputs and getting results somewhat akin to ¾" quality.

Generally, if you must perform interformat editing, start with the highest quality and allow the system to work it's way down to the lowest acceptable quality. If you're thinking about a particular approach, borrow some demonstration equipment and try it your way. I'm not convinced that there's a clear benefit to originating on a component format and editing on a heterodyne format–with but two exceptions.

If you are planning to slowly upgrade your facility from heterodyne to component, I would start the change in the acquisition format and progress through the system. This way, you can learn about component operations at a lower cost and with a minimum of sweat. (I'm sure that there are many diehard component users out there who will disagree, but you've already put in your learning curve in the school of hard knocks.) The other exception is in those cases when you can rent a component system on those occasions when you really need the high quality.

Now comes the quality factors associated with the VTR formats. *Generally* quality goes up as you progress from the heterodyne formats (VHS, Super VHS, 8mm, HI-8®, ¾", and ¾"-SP®), through analog component formats (M-II and BetaCam®), through direct analog composite formats (1"-C), through digital composite formats (D-2), and digital component formats (D-1). In this case, the quality of the method of processing, not the width of the tape is the deciding factor.

The output signal quality any formats will be higher with minimized time during which the luminance (Y—the black-and-white portion of the signal) is combined with the chrominance (C—the color portion of the signal). This is what the Y/C and Y/688 outputs are all about. In heterodyne VTRs, the Y and C signals are recorded in the same area of the tape at different frequencies. (Although there is a possibility of interference, it can be minimized with circuit and recording head design.) Where the Y/C and Y/688 outputs shine is provided by bypassing the circuit where the Y and C (or Y and up-converted 688) are recombined into a composite signal. Keeping the two signals separate at the output significantly improves picture quality. Keeping these same two signals separate throughout the production process can improve picture quality even more.

RULE #2

MINIMIZE THE AMOUNT OF TIME THAT LUMINANCE AND CHROMINANCE ARE COMBINED TOGETHER.

This is where the true "component systems" enter the picture (no pun intended). These systems require the C (same olde "chrominance") to be further broken down into its two basic parts ("R-Y" and "B-Y"—old systems once used "I" and "Q"). The Y, R-Y and B-Y are each recorded into their own individual area of the tape.

To properly use a component format, three separate channels (three cables, three connectors, etc.) are required to convey a single video signal. If the signals are to be transmitted, they must be combined into a composite signal before going out into the ether.

Again, we're following Rule #2 to the letter. It turns out that the two chrominance components interfere with each other, prompting

RULE #3

MAXIMUM PICTURE QUALITY CAN BE OBTAINED BY USING THE BASIC BUILDING BLOCKS OF THE VIDEO SIGNAL.

Every rule must have an exception. All this theory begins to fall apart as you approach the processing quality afforded by direct recording (used in 1"-C). The huge area of tape, expensive processing circuitry, and critically controlled VTR operation allows the playback a very high quality composite analog signal. In the 1"-C format, no picture information in the video signal is recorded separately.

All bets are off when you start talking about digital. In multiple generations of tape, nothing beats a digital recorder *in a digital environment* (digital switcher digital processing, etc.). When making multiple generations of digital recording in a conventional analog plant (like most of us currently have), it is not unusual for the analog-to-digital and digital-to-analog conversions to take their toll. Distortions from "quantizing errors" start creeping into the picture. Don't get me wrong—I think digital video is the best environment in which to work with multiple generations. It's just tough to get digital recording to live up to its potential in an analog plant. If you're thinking about digital, take a unit for a test drive under the anticipated operation conditions in your facility—then make your decision.

RULE #4

DIGITAL VIDEO RECORDING IS BETTER THAN ANALOG RECORDING *IF* THE ENTIRE SYSTEM IS DIGITAL.

The difference in picture quality from a D-1 facility (component digital) and a D-2 facility (composite digital) is akin to the differences between component and composite in the analog domain. It just drops back to Rule #3.

As you are probably painfully aware, the higher the picture quality, the higher the cost of the recorder. *But there's more to it than just that!* Look also at the cost of your support equipment. A "component plant" must have three-channel everything to derive the maximum benefit of a component format. A "digital plant" must have digital everything to derive the maximum benefit of a digital format. Guess what? The cost of the equipment to support the high quality formats also costs more. Which brings me to

RULE #5

THE HIGHER THE QUALITY OF A PRODUCT, THE HIGHER THE COST OF THE PRODUCT.

This is obviously a variation of "you get what you pay for." Don't think that a $3,000 "digital" camera is designed to give optimum picture quality through a $100,00 "digital" VTR (although some manufacturers may want you to think that way)!!! Of course, there are exceptions. Which brings me to

RULE #6

THERE ARE EXCEPTIONS TO EVERY RULE.

Ya better believe it, ya ha!

Q: What is "bias" in a tape recorder?

A: The "bias" in a tape recorder refers to the "carrier" signal onto which the baseband video or audio signal is "modulated."

The bias signal, before being modulated, is a pure sine wave. It has regular, periodic waves of voltage changes. Without modulation, these voltage changes reach a consistent peak voltage (maximum amplitude) without any change in the number of voltage excursions-per-second (frequency).

Frequency modulation changes the number of voltage excursions-per-second of the bias signal. Frequency modulation (FM) of the bias signal is the method usually used to record the video signal on tape.

Amplitude modulation changes the peak voltage attained by the bias signal. Amplitude modulation (AM) of the bias signal is the method usually used to record the audio signal on tape.

The amount and rate-of-change of the voltage or frequency variation is determined by the modulating signal. The higher the amplitude of the modulating signal, the greater the amount of change of the bias signal. The higher the frequency of the modulating signal, the greater the rate-of-change of the bias signal.

All this bias nonsense changes the operating environment in which the signals are being recorded. The modulated bias signal places the signal variations at a point where there will be a minimum of distortion encountered during the record and playback processes. The "bias" compensates for the response of the magnetic recording medium.

Regarding the question about bias that also appeared in the June article, Mr. Hurst points out that " . . . bias in a tape recorder is usually thought of as a summation process rather than a modulation process. Its purpose is to circumvent the natural horrible non-linearity of the tape's BH [Bias-Hysteresis] characteristic by rapidly and repeatedly exploring the BH characteristic, and then letting the final magnetization

settle down to the central value of the sum of the bias and the signal, which final value is the instantaneous value of the signal itself.

The FM system used in most video tape recorders is not a bias at all, but is the signal itself, no bias being used in the recording. The FM is produced in a modulation process not intrinsically different from any audio or video FM transmission system. We get away with recording this signal without bias because the BH non-linearities simply perform coring and clipping on the FM signal. A cored/clipped audio signal would be horribly distorted, of course, but a cored/clipped FM signal is just fine, since these actions do not change the information contained in the crossovers, which is the only place information is placed in an FM signal.

Q: ***I have been struggling with dubbing English translation onto ¾" masters where the total audio track is in Italian.... Is it preferable to leave the foreign language at a low level on one track so it can always be heard?... I feed one channel of the Italian into an audio mixer while the translation is being read into a second channel on the mixer. I have then fed a sum out back to the recorder and performed an audio dub; however I cannot raise the channel one audio without getting feedback. I receive a ¾" High-Band PAL master and dubs from this master are sharp and acceptable; however the picture quality from NTSC dubs is terrible. Would a format change be in line? —Kim Rowley, Focolare Movement, Washington, DC***

A: The issue about whether or not to leave the Italian at a low level is more of a subjective issue than a technical one. Keep in mind that if you don't mix the Italian back into the audio dub, it will not be heard at all (nor will any laughter or applause). I would think that some of the original sound would enhance the translation, possibly changing the relative level, when appropriate. This means mixing the two audio signals. The mixing can occur during the audio dubbing process or can occur during the playback of the tape.

You have chosen to mix during the audio dubbing process. This gives you control over the final aural scene. You can raise and lower the original sounds at will. The feedback loop problem, however, presents some interesting challenges. First, there must be a continuous loop in order to create the feedback. Possible areas in which the loop has been completed include a headphone worn by the interpreter and picked up by the interpreter's microphone (likely cause), from crosstalk within the mixer (unlikely cause), or from crosstalk within the VTR (unlikely cause). Not much that you can do about the last two possibilities, but they're unlikely causes anyway. Look at the arrangement through which the interpreter receives the original sound track. Unplug the interpreter headphones to see if the loop is broken and the feedback goes away. If that's your problem, try sealing the earphone cups better (or replacing them), move the interpreter's microphone further away from the headphones, and reducing the headphone volume to as low a level as usable.

There's one other possibility here. Make sure that you are using the opposite audio channels on the tape. Take your Italian feed from one channel on the tape and dub your new, mixed audio on the other channel on the tape. That might also be another source of feedback if not carefully controlled.

If your final audience were equipped with an audio mixer, they could adjust the balance between the Italian and English the way that they wanted. Unfortunately, most audiences won't know how to control the mix and all the effort will be for naught.

Now for the picture question. It sounds like that there are two critters at work here–noise and generation loss. First, the conversion from PAL to NTSC is a notoriously noisy endeavor. The higher the quality of the converter (and the higher the $$$), the less noise is induced into the converted signal. If there is an objectionable increase in noise in the picture, look for another converter.

The second issue is the age-old generation loss. If you receive a third generation PAL tape, send it through a standards converter and try to record the NTSC signal, you're right at the limit of useful generations of ¾" tapes. Yeah, a format change to something like ¾"-SP® *might* help, but I'm not sure it would be worth it. Borrow a demo ¾"-SP® deck from your equipment dealer and try it. It is my recommendation, however, that if you decide on a format change, move up to a format capable of a substantially greater number of generations (like BetaCam® or M-II.) That'll solve both the challenge of audio dubs on your "master" as well as the problems associated with multiple generations of tape. All it takes is a few million Lira.

Q: My folks in India own a multi-system AKAI (VX-9EGN) VHS VCR. It has nine world standards and plays NTSC American tapes. Just recently it has developed a snag when the NTSC tape is playing, the picture on the TV gets the jitters. It starts jumping–like you see three images of the person. Sound also garbles up. It doesn't run vertically so I imagine the Vertical Hold is OK. The majority of this occurrence is when they play tapes recorded in the EP or SLP modes. The SP almost always play perfect. The PAL system tapes also play perfect. Does this have something to do with speed or capstan adjustment? —VJ Sherring, Winfield, KS

A: This could me any of a number of things. Some possibilities include worn or dirty video heads, a dirty or hardened pinch roller, or even bad tapes recorded by a recorder that needs maintenance. I will assume that they have several sources of tapes (not just one), making the recorder problem less probable.

This leaves the heads. There is so little information recorded in the LP and EP modes that worn or dirty head problems show up sooner there than they do in SP mode. The amount of signal loss is there regardless of mode, but the initial lower signal level read off the tape in the other modes means that the same amount of loss will be noticed sooner in the EP and LP modes.

Try giving the heads a good cleaning with foam-tipped

swabs and head cleaner. If head cleaner is not available, try denatured alcohol (much more preferable than isopropyl "rubbing" alcohol). Touch the heads lightly and gently move the swab back-and-forth in the same direction as tape travel (parallel to the floor). Do *not* scrub the head up-and-down for it could break. Do *not* spray the head cleaner directly on the head–the thermal shock could shatter the head. After letting the heads dry thoroughly, start using the VCR in the normal fashion.

Sometimes you may need to repeat the process to get acceptable results. If, after three cleanings you still have the problem, consider replacing the video heads. In the USA, this is about a $150–$300 process on some machines, but you should check local rates in India; it may be less expensive to get a whole new VTR instead of trying to fix it.

Picture Displays

***Q:** How can I get rid of the dark line that appears between the green and magenta color bars on my color picture monitor?*

A: That dark stripe is perfectly normal on many color picture monitors. The phase change in the chrominance signal (representing color hue) is as bad as it gets—180 degrees. This instantaneous reversal of chrominance signal creates confusion in the color decoding circuits to create the illusion of the dark stripe. If you turn down the color, the dark stripe will disappear, proving that the effect is associated with chrominance (color components), not with luminance (black-and-white components).

***Q:** How can I get rid of dark horizontal bars that move up through my picture?*

A: Those dark bars are probably the result of a "ground loop." A ground loop occurs when there is more than one path to ground on a given cable. With shielded cable, the shield is usually connected directly to ground. If there is a problem with ground loops, try disconnecting one end of the shield from ground.

Disconnecting one end of the shield from ground is not a problem for balanced cables (where the shield encases two signal-carrying wires). For coaxial cable and other unbalanced signal cables, disconnecting the shield from ground will lose the reference voltage needed for proper signal transfer.

To provide only one path to ground for an unbalanced cable, a transformer is used to "isolate" the grounds that are necessary at each end of the cable. A transformer will pass the desired signal but will stop the effects of connected ground potentials. For audio work, these transformers are called "isolation transformers." For video work, these transformers are called "hum buck coils."

A word of warning here. DO NOT DEFEAT THE GROUND CONNECTION OF A POWER CORD!!! This will indeed lift the undesired ground, but it will also create a potentially lethal situation where the general public will read about your current event in tomorrow's paper.

Since I got off on the subject of ground loops, there is another possible reason why the dark bars are in the picture. The hum bars may be present if the power supply of one of the pieces of equipment has failed. The power supply may not be providing a true DC (direct, non-varying current) to the video amplifying and processing circuits. This is easily checked by a technician with an oscilloscope.

Q: ***How can you get the highest picture quality from existing video projectors?***

A: The easiest way is to keep the video signal components separated all the way from the camera to the projector. This bypasses the encoding and decoding of the signals and the distortion that those processes introduce.

As shown in Figure 7, start off with red, green, and blue video signals (RGB) from a video camera and other signal sources. (Remember you *can't* integrate the RGB from most computers with camera video–that computer stuff still isn't television-style video!!!!) Convert the RGB signals, as necessary, into Y (luminance), R-Y (Red minus luminance), and B-Y (Blue minus luminance) signals for recording on component video tape formats. Before projecting, convert the Y, R-Y, and B-Y signals back into RGB (unless your projector has Y, R-Y, and B-Y inputs).

The projected picture from the signals developed in this manner will knock your socks off! There is no "chroma crawl" to make the picture fuzzy. There is the ultimate sharpness of each of the projected red, green, and blue images. There isn't any interaction between color information and luminance (black-and-white) information.

On the downside, it takes three channels, three cables, and (possibly) transcoding equipment to make it all work. But boy does it work! The Bradley Center in Milwaukee uses this type of system for their twenty-foot General Electric multiple-light-valve replay screens. You can watch the hockey puck wobble as it flies across the rink!

Try it, you'll LOVE it!

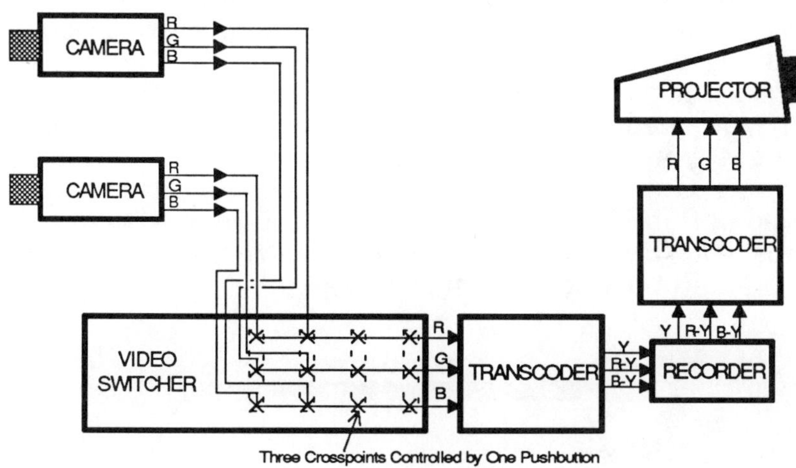

Figure 7
Component System Driving a Video Projector

***Q:** What kind of front-projection screen should I use with video projectors?*

A: The screens used for video projectors have exactly the same characteristics as screens used for other presentation media. Basically, the wider the area that can see a projected image (the "viewing angle" shown in Figure 8), the darker the image.

Some screens are designed to concentrate the light into a

narrower viewing angle. When a narrower viewing angle is used, the screen appears to increase the gain (AKA brilliance) of the projected image. There is only the projected light available to be spread. You can spread a dim image over a large area or you can spread a bright image over a small area.

With some video projectors, the projected light is so dim that a screen with a narrow viewing angle is almost mandatory. This, of course, is highly dependent on the projector and how hard you drive it.

For beaded screens, the recommended maximum viewing angle totals 50 degrees. Matte and lenticular screens can provide up to a 90-degree viewing angle. For any given screen material, check the manufacturer's specification sheet.

If you need more information about projection screens, I suggest you get a copy of "Kodak Projection Calculator and Seating Guide For Single- and Multi-Image Presentation" (Kodak Publication Number S-16) and ***images, images, images: The Book of Programmed Multi-Image Production*** (Kodak Publication Number S-12).

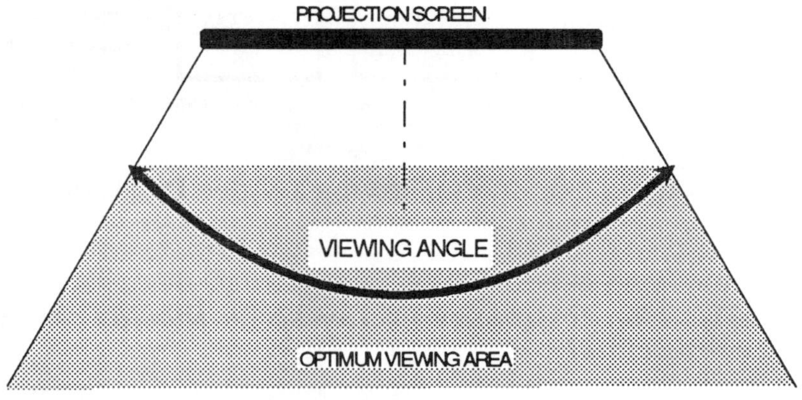

Figure 8
Screen Viewing Angle

***Q:* All the monitors in our editing suite have a horizontal dark band (about ¼" thick) rolling from top-to-bottom in an otherwise good picture. What can we do to eliminate the band? —Chris Mosio, Goal Productions, Inc., Altadena, California**

A: I called Chris on this one. Additional symptoms were discovered. The band was always there and moving at what was apparently a constant speed. Two editing benches with two different brands of editors were doing it. Grounding hadn't done anything. Running black burst to each piece of equipment in the system had been suggested (at approximately $500 cost).

The fact that the bar is moving at an apparently constant rate is an important clue. A moving picture distortion indicates that the interfering signal is not synchronous with the signal being displayed in the picture. If the distortion is stationary, the interference is synchronous with the displayed signal. If the interference is synchronous, the interfering signal is being created by equipment being driven by the same sync generator signal as the equipment creating the picture.

The other important fact is that the band is always present. If the distortion is always present, the source of the interference is always on. Equipment that is always on is equipment that is usually used in the system.

Chris tried turning off the status monitor (displaying the edit decision list–"EDL") and the band disappeared. The scan rate of a status monitor is usually different than standard video rate, so a bar would move through the picture.

It was discovered that the video monitors were sitting adjacent to the status monitor. Chris turned off the status monitor and the dark band went away.

What was happening was the deflection coil on the picture tube of the status monitor was sending out a magnetic field so strong that it was interfering with the magnetic field created by the deflection coil on the picture tube of the video monitor. Slipping a metal plate between the two monitors reduced the coupling between the deflection coils to the point where the interference went away.

$500 saved. APPLAUSE, APPLAUSE.

Q: What makes my picture monitor screen so dusty?

A: It was just a matter of electrostatic attraction. (Doesn't it sound romantic? Somebody ought to write a song.) There is a very high positive charge (as high as 50,000 volts) placed on a coating immediately behind the glass of the screen that in order to attract the negatively-charged electrons that start their trek from all the way back in the neck of the picture tube. Unfortunately, dust particles often become negatively charged and get attracted to the screen. Then, viola, you have a screen with so much dust that resolution begins to suffer.

There's another kind of particle that really yukkies up the screen–cigarette smoke. I have seen screens in smoke-filled rooms literally turn the entire picture brown from smoking byproducts. Smoke, like dust, is also composed of easily charged particles. Another reason to quit the filthy habit.

I haven't seen any special anti-static spray for picture monitors (or computer screens) that works. As soon as you turn the picture tube back on, the charge will be created and you have created a fatal attraction for the dust. (Kinda like a Bug Zapper!) Window cleaner and clean, soft rags are all you need to clean the tube. Try to remember to do it about once a month.

Q: What is the difference among "video," "data" and "multiscan" projectors?

A: Very little. A "video" projector is designed to accept signals from television equipment. This means that the projector must accept NTSC, PAL, SECAM or RGB signals. These signal types are analog with 1 volt peak-to-peak (from "low" sync tip voltage -to- the "high" maximum white picture detail voltage). It also means that the projector must be capable of scanning at the nominal rate for one or more of the major television standards. (In NTSC signals, the projector must be able to complete 59.94 vertical scans every second.) This means that there are the following:

$$59.94\ \frac{\text{fields}}{\text{second}} \times 525\ \frac{\text{horizontal scans}}{\text{frame}} \div 2\ \frac{\text{fields}}{\text{frame}}$$

$$= 15{,}734.264\ \frac{\text{horizontal scans}}{\text{second}}$$

A "data" projector and a "multiscan" projector today mean basically the same thing. A "data" projector will accept the output signals from a computer. These signals may be analog (at any of a number of voltage ranges) or digital (usually TTL–the 5 volt peak-to-peak signals used in "Transistor-Transistor Logic). A "data" projector is designed to be adaptable to many different computer scanning standards–there just ain't one kind of computer "video" signal. Some "data" projectors can display signals created at vertical scanning rates anywhere between 50 and 70 fields-per-second (and can be used as a "video" projector as well). Some "data" projectors can display signals created at horizontal scanning rates anywhere between 15,000 and 70,000 scans-per-second.

When I design a facility, I always install a "data" projector that can double as a "video" projector. You never know when the boss will walk in with spreadsheet data in hand.

Video Processors

Q: Should I buy an "image enhancer" or "processor" for the output of my VCR?

A: Probably not. I haven't seen one of these units for less than about $800 that is worth even considering. I've got one for which I paid $99.95. I bought it because it recreates every distortion I might want to demonstrate. This unit, and most of the low-cost units, actually add distortion to the picture and make it worse!

Q: What is "advanced sync?"

A: Advanced sync is a signal that is generated within some time base correctors (TBC). This sync signal is used to stabilize and synchronize the operation of a "source" video tape recorder.

Within a TBC, the video signal encounters a delay. The amount of delay is continuously varying, depending on the instantaneous time base stability of the incoming video signal.

A VTR can be locked to a synchronizing signal which has been advanced in time to anticipate the delay within a TBC. When the VTR locks to this signal, the advanced time in the sync signal is cancelled by the delay in the TBC. In this way, the video output of a TBC is synchronous with the timing reference supplied to the TBC (via genlock or sync lock inputs). If the TBC timing reference is house sync, the TBC video output is another synchronous source that can be used just like a camera or character generator.

Usually, TBCs with frame storage capability do not have an advanced sync output. This lock-up is not required with such a large amount of memory.

Q: What is time base error?

A: Conversion from daylight time to standard time and *vice versa*. Jet lag. Leap year. Time base error is any deviation from the amount of time it takes to complete one complete periodic event (daylight savings time conversion = one hour of time base error; jet lag time base error = the hour difference between the local clock and your biological clock; leap year = one day of time base error).

In television, time base error is based on the time spend during the scanning across one of those 525 (in North America) horizontal scans across the picture. If you dig out your copy of the FCC specifications or EIA RS-170 specifications and your handy-dandy Radio Shack calculator, you can calculate that the amount of time spent on one horizontal scan is 0.0000635 seconds. If the amount of time spent during *each* horizontal scan is not 0.0000635 seconds, you have time base error.

Any deviation in time base will show up in the picture as a variation in the size of the picture. Since you want a consistently sized picture, without unplanned zooms, you need to constantly maintain the time base.

Time base error occurs naturally in almost every piece of equipment; however, the video tape player introduces the largest amount of time base error. That's why you normally see Time Base Correctors (TBCs) connected to video tape players (or video tape recorders in the play mode). Even after going through a TBC, there is still a certain amount of "residual jitter" in the signal. (The amount of residual jitter is specified for each TBC and is one indication of the quality of the TBC. The lower the amount of residual jitter, the better the stability of the TBC.)

Q: Why do some TBCs have "advanced sync" and others do not?

A: "Advanced Sync" is a signal created within *some* TBCs to which the VTR to be corrected is locked. (The VTR must have a "SYNC IN" connector to take advantage of this arrangement.) On TBCs with limited correction range, the Advanced Sync signal locks VTR operation to the center of the range. If the correction range is not limited, or "infinite," there is no compelling reason to provide Advanced Sync to the VTR.

Q: Why do some TBCs have "VTR SC" and others do not?

A: Generally, a TBC with a "VTR SC" output signal is used to process the output from a heterodyne VTR (like ¾", VHS, S-VHS, BetaMax®, 8mm, and Hi-8®) to create broadcast-quality video signals. *Some* heterodyne VTRs (by all means *not most* of heterodyne VTRs) have an "SC IN" connector to accept the VTR SC signal. When an appropriately equipped VTR is connected to an appropriately equipped TBC, the TBC may be operated in the "direct" mode so that the corrected output meets FCC regulations for subcarrier-to-horizontal sync (SC/H) phasing requirements. In this mode the playback color correction circuit in the VTR is slaved to the subcarrier reference signal created by the TBC.

If a TBC does not have a "VTR SC" output *and* there is no other provision for reestablishing SC/H phasing (sometimes in the TBC output processing circuitry), the corrected video signal will not meet broadcast standards.

Q: What is an "infinite window" on a TBC?

A: The "Window" of a TBC is the amount of correction that the TBC can perform. An "Infinite Window" means that there is so much memory in the box that there is no practical limit to the amount of correction because it can replace unwanted frames (or portions of those frames) with good information with little impact on the picture.

Q: What is a synchronizer?

A: In video, a synchronizer is a black box that converts incompatible video signals into video signals that can be used together. A synchronizer alters the timing of video signal sync and picture elements.

In audio parlance, a synchronizer locks the operation of one device (frequently a multi-track audio recorder) with another device (frequently a video recorder). The synchronizer slaves the rate of playback from the audio recorder to match the video recorder. This eliminates problems with lip synch and other audio-to-video time distortions.

Audio synchronizers often reference SMPTE Time Code signals or MIDI Code signals for their operation.

Q: What are the differences between a "frame synchronizer and a "TBC?"

A: The basic difference is that a Frame Synchronizer expects the input video signal to have a stable time base. A TBC expects its input video signal to need correction of time on a line-by-line basis.

Many boxes have a TBC/Frame Sync switch that will allow both functions in the same piece of equipment. In general, when connected to correct the video signal from a VTR (when a significant amount of time base error is expected), the switch should be in the TBC position. When the equipment is not used to correct the video signal from a VTR (where a significant amount of time base error is *not* expected), the Frame Synchronizer mode should be selected.

On some pieces of equipment, if the TBC/Frame Sync switch is left in the TBC position and you are using the equipment for a Frame Synchronizer, the equipment may actually induce a small amount of time base error into the video signal.

Test Equipment

Q: I don't think that my waveform monitor is reading correctly; what could be wrong?

A: The first thing to look for is the calibration of the waveform monitor. Switch to "CAL" and make sure that the displayed square wave is exactly 140 IRE. (This assumes that the square wave calibration signal has recently been adjusted to the voltage reference of the National Bureau of Standards.)

Once a waveform monitor is properly calibrated, problems may occur because of inaccuracy of the termination of the signal being measured. If the termination load (in Ohms) is too low, the waveform monitor will read too high. If the termination load is too high, the waveform monitor will read too low.

Precision terminations are usually supplied with waveform monitors to get around this termination quality problem. These precision terminations are expensive (frequently as much as $20–$25) but are necessary to get accurate measurements from a piece of test equipment.

Q: What is displayed when a waveform monitor is adjusted to show "2H?"

A: When most low-cost waveform monitors are adjusted to display a "2H" display of the video signal, they actually show the entirety of the video signal two horizontal scans at-a-time. The two horizontal scans side-by-side *do not* represent two fields. Instead, the displayed line number sequence goes something like:

1 — 3
5 — 7
...
241 — 243
...
523 — 525
2 — 4
6 — 8
...
242 — 244
...
246 — 248
260 — 262

(Remember that horizontal scans are numbered from top-to-bottom as they appear in a picture. A picture consists of two vertical scans or "fields." The scanning sequence in a given field is either even-numbered lines or odd-numbered lines.)

In the "2V" display mode, the two fields are indeed side-by-side and separated by the vertical sync pulse.

Q: What test equipment do I need?

A: Depends on what you want to do. For most mid-sized corporate production activities, I recommend the following video equipment to support production operations (in descending order of importance):

- Stable moderate-quality picture monitor ($1500–$2500);
- Waveform Monitor ($1200–$2500);
- Color Bar Generator ($1200–$3500); and,
- Vectorscope ($1500–$3500—may be combined with Waveform Monitor without any sacrifice in quality or operational convenience).

The operational video test equipment should be complemented with the following operational audio test equipment:

- Moderate-quality audio speakers ($500–$1500, each);
- Moderate-quality audio power amplifier ($500–$1000);
- Tone Generator ($100–$500); and,
- Voltmeter designed for frequently encountered audio levels and calibrated in Volts and dB ($500–$2000).

There's another type of test equipment you will need if you actually want to service the equipment. In that case, you need:

- Service manuals (*not* operator manuals) for each piece of equipment that you are attempting to support (usually $100–$200 each);
- A good volt-ohm-milliamp meter ($50–$300);
- A dual-trace general purpose oscilloscope ($600–$4000);
- A frequency counter ($400–$2000); and,
- Any specialized equipment required in the SERVICE manuals.

With careful selection of test equipment for servicing, both video and audio equipment requirements can usually be economically fulfilled.

Fractured Definition:

---

mul·ti·burst \'məl-ti-bərst\ *v* **1 :** Roseanne Barr in petite pantyhose **2** state of co-producers who are over budget **3** emanation from the rusty pipe over your most expensive piece of equipment *n* **1 :** a test signal used to measure video frequency response

I've had some second thoughts about this next question about measuring resolution with a waveform monitor. A waveform monitor cannot measure the resolution of a picture monitor because resolution is a visual constraint, not an electrical constraint. The waveform monitor can be used to measure resolution that is limited by an electrical restraint imposed as the signal goes through various circuitry.

Q: Can you measure resolution with a waveform monitor?

A: Yes, depending on the waveform monitor. Set up your resolution or your registration test chart, as usual. (Make sure to calibrate the chart to the electronic television system by adjusting picture size so that the wedges on the edges of the chart are on the edges of the scanned picture – as displayed on an underscan monitor.)

Adjust the waveform monitor to display a single horizontal scan. Display the video signal output from the "MONITOR" or "PICTURE" connector on the back of the waveform monitor. Yes, this procedure *requires* a waveform monitor with single-line display capability.

As shown in Figure 9, move the viewed scan using the waveform monitor "LINE SELECTOR" control until the voltage spikes representing the converging lines in the middle of the chart is 50% of the maximum voltage spikes anywhere else on the chart.

On the picture monitor, look at the position of the highlighted scan path. The number of TVL (TeleVision Lines) of resolution on the horizontal axis of the picture can be extrapolated from the position of the highlighted scan path relative to the numbers printed on the chart adjacent to the converging lines.

What we're actually measuring here is the optical and electrical response of a camera to the black-to-white transitions, as well as some of the electronics. Camera pick-up device response, or "depth of modulation," is also included in your result.

You cannot measure the limiting resolution of a picture monitor with a waveform monitor.

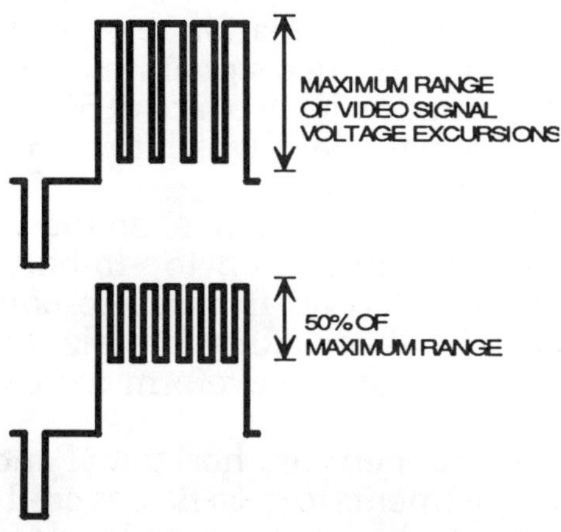

Figure 9
Measuring Resolution with a Waveform Monitor

Q: Why can't you measure the resolution of a picture monitor with a waveform monitor? —Sam Stalos, AV Video Magazine

A: You can only measure the resolution of a *signal* with a waveform monitor. You can measure the resolution of a *display* only with your eyes viewing the picture.

Q: How do I use a waveform monitor in setting up lighting?

A: This one's going to take experimentation and patience, depending on the results you are trying to achieve.

First you can adjust the illumination level to be even across an entire scene. Display "1H" (one horizontal scan) or "2H" (horizontal scans) of the video signal on the waveform monitor with the luminance filter (may be called "IRE" or "Low Pass") in circuit. The illumination level is even from left-to-right in the scene is the video signal waveform is horizontal. An average illumination level of about 50 IRE will generally give an acceptable base with which to compare other scene detail illumination levels.

Display "2V" (two vertical scans) on the waveform monitor. Adjust the illumination from top-to-bottom of the scene. The illumination level is even from top-to-bottom of the scene when the video signal waveform is horizontal. (Remember that, in this display, left of waveform corresponds to top of scene.)

Go back-and-forth between horizontal and vertical scene illumination adjustments to get a flat, overall scene illumination. Now increase the illumination levels in those areas of the picture that are supposed to be closer to the viewer. Get 'em up to about 80 IRE. (Remember that the luminance level of Caucasian skin is designed to be 77 IRE in the NTSC television system.)

Now drop the illumination of the distant objects down to about 30 IRE.

The result of this procedure is a base of illumination from which the individual scene details may be modeled. The scene will not be interesting until you make the final adjustments with a fine eye, viewing a properly adjusted picture monitor and being interpreted by the computer that resides between a pair of human ears. Don't expect the waveform monitor to do too much!!

If you don't have a waveform monitor at your disposal, try lighting the scene with one eye closed. Remember that television is a two-dimensional medium without any real depth. To eliminate the extraneous depth information that is provided by normal binocular vision, close one eye. You'll be surprised how much easier it makes things.

Q: What is the "1 Volt / 4 Volt" switch position on my waveform monitor used for?

A: Measuring signals with different amplitudes. The "4 Volt" position means that a display of a signal of 1 Volt peak-to-peak (bottom-to-top) will have a vertical size of 140 Units (from the -40 Unit Line -to- the +100 Unit Line on the graticule).

Use the "4 Volt" position to display and measure sync signals. This means that a display of a signal of 4 Volts peak-to-peak will have a vertical size of 140 Units. Remember from the previous question that most of the signals used in a sync lock system have peak-to-peak amplitudes of 4 Volts; one of those signals would be displayed as 140 Units in the "4 Volt" switch position. A subcarrier signal (with 2 Volts peak-to-peak) will be displayed as 70 Units in the "4 Volt" switch position—this would be an easy way to see what a "sine wave" looks like (hint, hint).

When looking at sync pulses, you should be making sure that they are rectangular (with sharp corners and fairly straight edges). If they are not rectangular, they could create an unstable picture that jitters. If they are not rectangular, suspect a bad cable, bad connector, too long a cable, a problem in the sync generator circuit that is generating the signal, or a problem in the Waveform Monitor circuits.

Make sure that the sync pulses are not varying in horizontal position on the display. Signals moving in the horizontal axis of the Waveform Monitor are varying in their time. If your sync pulses vary in time, everything varies in time. *This is a must fix situation* (as long as you're not looking at the output of a Video Tape Recorder without a Time Base Corrector). If the pulses are moving, either the waveform monitor "trigger" circuit has a problem, or the sync generator has a problem.

Q: What is a "routing switcher?"

A: A routing switcher is a matrix of transistor switches in an array that allows connection of one of several inputs to one or more outputs.

Routing switchers are normally described as the maximum number of independent input signals "-by-" the number of individual output signals. A 10-by-2 (10 X 2) switcher will accept ten independent input signals from which two independent output signals are available. As shown in Figure 10, each input signal is connected to an "input bus" that may be connected to any "output bus." Only one selected input bus will be connected to any selected output bus at any given time.

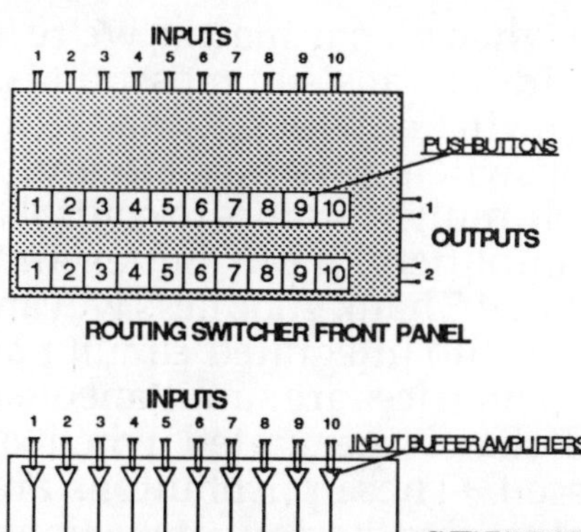

Figure 10
Routing Switcher

To overcome the losses associated with the input bus–transistor switch–output bus, an output amplifier circuit is usually installed between the output bus and the output connectors. It is not unusual for the output amplifier circuit to provide two identical output signals from the one signal on the output bus.

Routing switcher input buses often have a "buffer amplifier" to minimize distortions to the input signal. The buffer amplifier minimizes the effect on the input signal as various combinations of output buses are switched onto the input bus.

Each type of signal must be independently switched. An "audio-follow-video" (AFV) switcher has two transistor switch matrices. One matrix switches the video signals. The remaining matrix switches the audio signals. (A stereo AFV switcher will have three switching matrices–one for video and one for each of the two stereo audio signals. A stereo AFV switcher for component video will have five switching matrices–two for the audio and three for the component video. If you add tally lights to the switcher, that's frequently one more matrix!)

Keep in mind that when we say matrix, we're talking about the number of individual transistor crosspoints. A 10-by-10 stereo AFV routing switcher with tally has 10 × 10 × 4 = 400 individual transistor switching points. Now add another 800 transistor monostable multivibrators to control the switches. Now add output amplifiers, input buffer amplifiers, and power supplies. Whew! Thank goodness we can pack a large number of transistors into integrated circuit packages!

These independent matrices are simultaneously controlled by a digital signal that is generated when one common pushbutton is pressed. These pushbuttons are usually arranged in horizontal rows to display connections to an output bus. Connecting Input Bus #1 to Output Bus #2 is simply a matter of pressing the leftmost button (#1) on the second row of buttons.

Keep in mind that these rows of buttons may be sliced into individual rows and installed throughout the facility. Just keep in mind that you are controlling one output bus at a time and you will be OK.

Now, let's replace the pushbuttons with a digital command panel. By entering the numbers into a keypad, thumbswitch, computer, or whatever, you can connect Input Bus #1 to Output Bus #2. As far as the switcher matrix operation is concerned, this is *exactly* the same thing as pressing the leftmost button on the second row of buttons. Sometimes it's just easier to change the way the commands are entered.

It is interesting to note exactly what happens when you push the button on a routing switcher. First, any existing connection on the controlled output bus must be cleared. Then the new connection may be made. In this way, there is no overlap of signal to create a flash or other glitch in the picture.

Now that we understand the basics, let's look at how routing switchers are used. I use routing switchers when rapid or repeated switching among several sources is necessary during the normal course of facility operations. For example, I specify a routing switcher if I am attempting to use a single VTR for production activities from two studios, as an editing VTR, as a backup VTR, as a source for production, etc. In this case the routing switcher augments (but *never* replaces) a carefully designed hard patching system. In my opinion, patch panels still must be installed in case the routing switcher fails (which, according to Schultz's Laws, MUST happen).

In those applications where the routing switcher must "hot switch" (switch the signals in the active program instead of preview or other "off-line" function), the routing switcher should be capable of vertical interval operation. This allows the switch to be accomplished at a time when it will be invisible to the viewer. Of course, to use this function, each of the video signals must be properly timed and synchronized at the input to the routing switcher. Vertical interval switching is useless if the video signals applied to the switcher are not synchronous.

What should you look for when evaluating a routing switcher for your applications? First determine need. How many inputs and outputs do you need to control? Always overestimate potential need if you can afford it.

Now let's look at specifications. Crosstalk is the amount of signal that leaks from one bus to another. Crosstalk is usually specified in dB. The higher the number, the better.

Differential Gain and Differential Phase specifications describe the accuracy of the color signal after it goes through a switcher. Differential Gain is usually specified as a percentage and Differential Phase is usually specified in degrees. The lower the number in both cases, the better.

Bandpass is a function of how accurately *all* of the signal squeaks through the switcher. Bandpass is expressed as a range of numbers measured in Hertz (Hz). The wider the range of numbers, the better. (Remember that 1,000,000 Hz = 1,000 kHz = 1 MHz.)

Noise and hum are interfering electrical signals. The extent to which these signals interfere with the proper operation of a routing switcher is expressed as the "Signal-to-Noise Ratio" and measured in dB. The higher the number the better.

Q: What is Ultimatte®?

A: The Ultimatte® process is what I consider to be an enhanced chroma key. (Remember that chroma key "looks for" the picture areas in which a particular color appears and replaces those areas with another picture.) Ultimatte® takes this process one step further by looking for variations in the luminance of the selected color. These variations in selected color luminance are then imposed on the replacement picture (in an amount proportional to the original luminance change).

Looking at the usual "weatherman" example, assume that the weatherman is pointing to the blue set (that will be replaced by the weather map). The shadow from this hand and arm fall onto the blue set as he points. Ultimatte® takes this shadow (lower luminance) and transfers it to the inserted picture (the weather map). In the finished picture, the weatherman appears to be pointing to a real map, shadows and all.

There's a whole bunch more that the Ultimatte® process can add to many productions–it ain't just a matter of shadows. For more information, Ultimatte® has a good series of application notes. Telephone (818) 345-5525.

Q: What can I do electronically to modify our Sansui A/V 99 processor to do what Ultimatte® does?? We want to put live people in front of a computer-made set, cheap, for plays, etc. If you don't try, you can't fail! —Marvin Kantorowitz, EDCOM, Plainview, New York

A: There are several challenges here. Modification of a piece of equipment may intuitively be less expensive than buying the genuine article, but if you include time and materials associated with the project, the costs are usually significantly more that purchasing new and appropriate equipment outright. I've got an electrical engineering degree and I've done my fair share of "modifications," but they have always taken at least ten times more time and material than I ever estimated. It may be fun–until you must do it on a deadline.

My suggestion in your case is to contact some used equipment dealers to see if you can get an operable cheap chroma keyer or production switcher—I don't think you're going to find Ultimatte® capability. This solution, to me, will bypass many of the headaches of trying to modify a piece of gear designed for consumer operation.

If you must try the modification route, study some of the texts about how "keying" systems work and study the basic electronics texts about how transistor "switches" work. Your copy of ***Mastering Television Technology: A Cure for the Common Video*** will help on the former reference. There are lots of books about the latter in almost any library. You must have the schematic of the processor and find the point where the "keying" would best be inserted into the processor circuitry. Once you find that point, then you need to design a keying circuit. Then you breadboard that circuit and try to make it work (circuits *never* end in the same configuration as they began). Be ready to cut circuit board traces and install a custom made "piggyback board" with the components necessary for proper keying operation. Then you've got to install the controls on the front of the processor and connectors on the back of the processor and make sure that the power consumption of all this new stuff doesn't outstrip the capability of the original power supply. Once all of that is through, then align the operation of the new circuit and the old circuit to optimize the quality of the signal as passed through the equipment. Then you're through. Forget any technical support–you've violated all warranties, technical specifications, or hope of having a nice, clean panel layout.

Sometimes it's worth all the trouble it takes to modify equipment. Most of the time, it does not warrant all the frustration.

The second challenge is the interface between your computer and the television system. That's sometimes a difficult and expensive project.

Up from the depths of depravity:

HDTV = High Dollar TeleVision
Sync Generator = A porcelain and faucet manufacturer
Proc Amp = A device that makes it easy to do a prostate exam
ADO = A commotion, a stir (as in "Much ADO about nothing." Thanks, Mark)
DVE = Dumb Video Enterprise (I knew it was a bad idea to begin with!)
TBC = Too Bad Charlie (Good taste doesn't count any more!)
SEG = Something Entirely Gratuitous (Why is this training tape four hours long?)
VTR = Very Terse Response (Does the boss *really* not like it?)
VCR = Very Candid Response (Now I *know* he doesn't like it!)
Drop Frame = Ex art gallery employee
Non-Drop Frame = Soon-to-be Ex art gallery employee
SCH = Somebody Can Help
Non-SCH = Nobody Can Help
FCC = Friendly, Cordial, Communicative (Rrriight! Like "I'm from the FCC and I'm here to help you.")

Will somebody please come get me out of this mess!!

Q: What is the simplest and least expensive way to electronically flip a video image; how does one make a left-to-right image read the other way around? –Jorge A. Gonzalez, Brooklyn, NY

A: With a mirror between the picture and the viewer (or camera). If you're trying to record the picture, genlock the camera to the picture being displayed to reduce some of the dark bands moving through the picture.

Another approach–when shooting the original material, you could shoot into angled mirror(s) between the talent and the lens. A more elegant approach (I've spent too much time around architects when I talk about "elegant approaches"), use a tube-type black-and-white camera and reverse the horizontal or vertical deflection leads to the yoke around the tube. If you've got a multi-tube color camera, you've got to reverse the leads on each yoke and register the camera (ugh!) and register it again after normalizing the leads (UGH!). Knowing how they work, I don't think I'd even try it with a single-tube color camera unless I want a weird color effect.

A third approach (the easiest, highest quality and most expensive) is to use some sort to digital video effects (DVE) processing equipment. (You don't *have* to own it!!)

Audio Equipment

Q: What is meant by input/output levels of -10 dBm, 0 dBm, +4 dBm, etc. If it is advantageous to have these levels balanced throughout a system, how is this accomplished? — Wallace E. Steinbrecher, U.S. Border Patrol, Glynco, GA

A: First, I want to make sure that we realize that we are talking two different animals here—level and balance configuration. There is no direct correlation between these two concepts. First, let's take on the level question.

In audio, we normally measure signal level in terms of power as opposed to voltage or current. An alternative way of expressing signal power is with a logarithmic ratio with a "standard" signal power. This is where we come up with the dB (deciBel) ratio where

$$\text{dB} = 10 \times \log\left(\frac{\text{Signal A Power}}{\text{Reference Signal Power}}\right)$$

The same formula rearranged to derive the Signal "A" Power from a ratio expressed in dB gives us:

$$\text{Signal ``A'' Power} = (\text{``Reference'' Signal Power}) \times 10^{\left(\frac{\text{dB Ratio}}{10}\right)}$$

For the term "dBm," 0.001 Watts (1 milliWatt) is the "Reference" Signal Power. That's what the "m" in "dBm" means. If you dig out your handy-dandy Radio Shack scientific calculator and do the math, you'll find that the following signal levels result:

-10 dBm	=	0.0001 Watt (0.1 milliWatt)
0 dBm	=	0.001 Watt (1 milliWatt–same as "reference" power)
+4 dBm	=	0.0025 Watts (2.5 milliWatts)

Contrary to what is popular opinion, they didn't do it this way just to confuse you. It turns out that the ear (and the eye) sense in a logarithmic fashion. A doubling of sound power does not double the perceived volume. Blame it all on the guy in charge of the committee that designed our basic senses.

Now, let's attack the balance configuration part of the question. There are two basic ways to send a signal down a cable–balanced and unbalanced. Thanks to my internal wiring, I am an expert on the unbalanced configuration.

Balanced is better. Balanced is a little more expensive. If I were to be reincarnated as an audio system, I would much prefer to be balanced. That way I would be more immune to interfering signals than my unbalanced brethren.

Unfortunately, all the equipment manufacturers haven't heard that balanced audio is better audio. To correct their product deficiencies (hint, hint) there is an easy "fix"—the "balun" (BALanced/UNbalanced) transformer. A balun is a cheap and efficient way of connecting between a balanced and unbalanced signal conditions (either direction). Baluns are available from several manufacturers.

If you don't use baluns in a system to interconnect balanced and unbalanced signals, the background noise (hiss) will increase. A couple of mismatched balance configurations will not matter, but several mismatches system-wide may become objectionable.

Q: What is 0 VU?

A: Anything you want it to be. It's a reference Volume Unit. Everything is referenced to the "0."

With some mixers, 0 VU represents 1 milliWatt into 600 Ohms (0 dBm). On some audio tape recorders (and audio channels of video tape recorders) it represents the maximum input signal level for which the signal processing circuitry was designed. On a transmitter, it frequently represents the maximum amount of modulating signal that can be used without creating distortion products.

Q: What is the preferred pin basing on balanced audio connectors?

A: The method I prefer is

Pin 1 – Ground
Pin 2 – Signal
Pin 3 – Signal

Remember that in a truly balanced system, the signal is divided equally between two wires. The phase on one wire is 180 degrees out-of-phase with the other wire. The input circuitry is designed to look for the difference in voltage between the two wires. For this reason there is *usually* no preference between the two signal-carrying wires.

Q: I'm interested in buying a demagnetizer for my audio heads in various tape decks. I have noticed several different ones varying in price from $4.60 up to $45.00. Are there any differences among these? —Brian Schaff, Miller-Dwan Medical Center, Duluth, Minnesota

A: There two basic variables in the quality of demagnetizers. First, what is the strength of the magnetic field created by the demagnetizer? Second, what is the quality of construction.

The strength of the magnetic field is dependent on the number of turns in the coil of wire in the demagnetizer as well as the diameter of the wire in the coil. It is also dependent on the construction of the stinger that is used to concentrate the magnetic field.

In construction, the coil will be less likely to burn out if the diameter of the wire is larger. The insulation (usually just a layer of lacquer) applied to the wire before it is wound into a coil must be thick enough to eliminate the possibility of a short with continued use. Remember that anytime the demagnetizer is used, it will heat up and cause one coil to chafe against the adjacent coil. This chafing will eventually erode the insulation and create a short circuit.

When purchasing a demagnetizer, it's going to be tough to determine the strength of the magnetic field, the size of the wire, and the quality of insulation. This information is rarely printed somewhere. Indeed, this is one of those marketeers' delights where more expensive is not necessarily better.

I'd look for some sort of thermal shut-off like a circuit breaker or thermister. I'd also look for approval by a testing agency like Underwriters' Laboratories (UL).

Some things to remember about using demagnetizers:

1. Never demagnetize a video head. It's not needed and may shatter the head.
2. Never touch any metal on the demagnetizer to the head you are demagnetizing.
3. Turn on the demagnetizer at least three feet away from the tape transport, slowly approach the heads you are trying to demagnetize, move the demagnetizer up and down and around the head your are trying to demagnetize (all the while chanting some unrecognizable tune), then slowly remove the demagnetizer to about three feet before you turn it off. (Your boss may be unimpressed with your chanting ability, but he will think you aren't doing anything if you don't chant.)

When do you demagnetize? My philosophy is to demagnetize after you start hearing an increase in the background hiss in the audio. Some books I have read that address recording studio applications recommend doing it every morning. (You probably have better things to do every morning!)

Q: What is impedance?

A: Impedance is opposition to the flow of varying ("AC") electricity. (Resistance is the opposition to the flow of nonvarying "DC" electricity.) There is a phase consideration (minute timing differences between the maximum voltage and the minimum voltage of a signal that is rapidly varying) in impedance that is not present in resistance.

Impedance is critical in all of electronics because of this phase consideration. In video, impedance distortions create losses in resolution (picture detail). If impedance is not maintained correctly through a video system, you can easily loose a hundred or so of those expensive "lines of resolution" for want of a $.39 part! Make sure that the impedances are matched at the ends of the signal path–you can't simply put a termination on a coaxial "T" anywhere in the circuit–it must be at the physical end of the signal path.

In audio systems, impedances among the various pieces of equipment must be matched in order to minimize the distortion to sibilant sounds (like "ssss," "sssc," and "kk"). You've heard this type of distortion on a poorly tuned radio. Match the distortion and, viola, you get rid of the distorted sounds.

When it comes to matching impedances among various pieces of equipment, the video folks have their act together. Almost all video equipment has a characteristic input, output, and cable impedance of 75 Ohms. Audio folks run in circles chasing 4 Ohm, 8 Ohm, 110 Ohm, 150 Ohm, 600 Ohm, and 10,000 Ohm impedances (among *many* others).

Because of the lower signal frequencies involved, audio signals *appear* to suffer less distortion than video signals when impedances are not matched. Don't let this lure you into a sense of complacency, however, because the distortion effects are cumulative as you go through a system. If you don't pay attention to the matching of impedances, at the end of the line, you may wind up with distortion products created from distortion products created from distortion products created from distortion products created from distortion products . . . It eventually becomes almost an exponentially increasing number of distortion products if you keep this up long enough.

It's cheap and easy enough to match these impedances in the first place. It will take a handful of resistors or transformers and a little knowledge. The ***Handbook for Sound Engineers: The New Audio Cyclopedia*** has a pretty good chapter on the subject. (Look in the index under "Impedance Matching.")

***Q:** What is the best way to connect speaker level outputs to line level inputs? –Conrad Shull, Williamsport Hospital & Medical Center, Williamsport, Pennsylvania*

A: Verrrry carefully. The signal power emanating from a speaker level output can be up to a couple of hundred watts or more. The signal power expected by a line level input is about a millionth of a watt. For this reason, leave the speaker in-circuit to dissipate most of the power. In essence, you will "bridge" (connect in parallel) across the speaker terminals; BUT, before you do anything, turn the amplifier power off!

The second problem that you'll run into will be matching the impedance to minimize distortion. A typical speaker impedance is 4, 8, or 16 Ohms. A line level input normally has a 600 Ohm or 10,000 Ohm impedance. In the case of a line input with 10,000 Ohm impedance, connect a 10,000 Ohm (10k) resistor in one of the two conductors connecting the speaker terminals with the line level input, as shown in Figure 11. In the case of a 600 Ohm input, place a 600 Ohm resistor between the two conductors between the 10,000 Ohm resistor and the input.

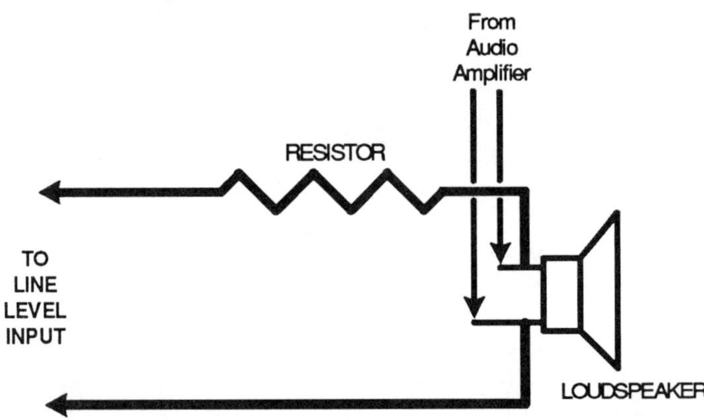

Figure 11
Connecting a Line Level Input to a Speaker Level Signal

Now reduce the volume to minimum. Then turn on the amplifier. Then turn on the speaker (if it has a separate switch). Then *SLOWLY* turn up the volume until you get an acceptable level *AT THE INPUT CIRCUIT*. You can determine if the level is acceptable by watching a VU meter on the input circuit or by listening to the output of the equipment into which you are sending the signal. The volume of sound produced by the speakers is not important in this discussion—just don't make it too loud–the neighbors and/or your speakers will complain. (Normally the volume is so low that you will just barely be able to hear anything from the loudspeaker.)

For stereo-to-mono mixdown, I'm not sure what all the ramifications are. I'd probably use a mixer with line level inputs to do it right. Each line level mixer input would be individually connected to a speaker.

As you go through this exercise, make sure you are bridging the entire signal. If there is a "crossover network" driving the speakers, bridge the input to the crossover. The outputs of the crossover network contain frequency-selected portions of the entire signal. (Crossovers are frequently installed within the speaker cabinet—connect across the terminals on the back of the cabinet and you're OK.)

One last bit of caution. If you're digging into a consumer piece of equipment, it may have a "hot chassis" where one side of the power (instead of ground) is directly connected to the chassis. (It's cheaper and deadlier to make that way!) You're usually OK on the speaker, but beware the chassis in that case.

Q: What's an octave?

A: An octave is a doubling of frequencies. Two tones separated by an octave sound similar, but the intermediate tones are not separated by equal frequency differences.

Q: What is phantom power?

A: If a microphone that requires power for proper operation does not have a built-in power supply or battery, the microphone is being operated with phantom power. Phantom power is sent from the power supply (that may be in an audio mixer) to the microphone via the same wires that carry the microphone audio signal. The audio signal is separated from the power signal via transformer coupling. A transformer will block the non-varying "DC" component of the signal but will pass the varying audio "AC" component of the signal.

Most modern phantom power systems operate at 24VDC although some older systems operate at 48VDC. The 48VDC systems will not normally damage microphones of modern design.

Radio Frequency (R.F.) Equipment

Q: What is a modulator?

A: A modulator is a miniature transmitter. Indeed every "transmitter" has a "modulator" circuit. This circuit takes the "baseband" video and audio signals (the "raw" signals from cameras and microphones) and place them in a specified bandwidth "channel." In television modulators, the video is impressed into a "visual" carrier and the audio signal is impressed into a separate "aural" carrier.

The visual carrier is a signal with a frequency that is specified to be within a particular range of frequencies called a "channel." The amplitude of this "very high frequency" (VHF) or "ultra high frequency" (UHF) carrier normally varies so that the peaks and the valleys stay at the same voltages. When a video signal is "amplitude modulated" onto the visual carrier, the voltages of the peaks and valleys vary to reflect the shape of the applied baseband video signal.

At the other end of the transmission system a DEmodulator (often called a "tuner") essentially strips the baseband video signal from carrier. The output from the demodulator is a pretty accurate representation of the original baseband video signal.

Why go through all of this nonsense? First, it's much easier to send such a signal through the air. Second, several modulated visual carriers of different frequencies can go through the same space or cable without significant interference. Third, these carriers can go in different directions through a space or cable.

What's the down side? Distortion. All of this modulation/ demodulation processing can really make the signal degrade significantly. Have you ever looked at the difference between the R.F. ("Radio Frequency) output signal and the video output signal from your home VCR? You will see a significant degradation to the video signal as it goes through all of these processes. The distortion show up as reduction of "resolution" (that is frequently called "fuzzier." There will also be an increase in the amount of noise or "snow" in the picture.

That's video–how about audio? Well, TV channel audio modulation is the same idea but different. The high frequency carrier onto which the baseband audio signal is modulated is called the "aural" carrier. In North America, the frequency of the aural carrier *without any modulation applied* is 4,500,000 cycles-per-second (4.5 MHz) higher than its associated visual carrier. The baseband audio signal is "frequency modulated" onto the aural carrier. This means that the *amplitude* of the applied baseband audio signal is represented by the amount of *frequency change* (in cycles of voltage-per-second) of the aural carrier. The frequencies in the audio signal shows up as the amount of "deviation" in the aural carrier from the "at rest" frequency.

Higher Audio Volume = Greater Aural Carrier Frequency Change

The *frequency* ("pitch") of the applied audio signal shows up as the *rate-of-change* of frequency of the aural carrier.

Cycles-per-second of baseband audio = Rate-of-change of the aural carrier.

Distortions to the audio signal as a result of the modulation of the aural carrier usually shows up as excessive background noise and excessive stress on the consonant sounds.

Notice that the amplitude of the aural carrier is not terribly important. (Yeah, it's got to be strong enough to be detected, but amplitude will not really affect the amount of aural carrier deviation or the rate-of-change of the deviation aural carrier.)

Amplitude of the visual carrier is critical. Amplitude of the aural carrier is not as critical. Distortions that affect amplitude will show up much faster in the picture than in the associated sound. Lightning is a prime example. Lightning discharges from a thunderstorm will readily show up in the picture (it creates noisy amplitude spikes in the same area through which the signal must pass), but will have minimal effect on the sound.

Here's some homework for you ...

The next time a thunderstorm approaches your house, look at the off-the-air signals (not via cable–they may be pulling some funnies on you). You will notice that the effects of the lighting strikes will be much more pronounced in the picture than in the sound.

Now look at the effect of lightning on the low VHF (Channels 4–6), the high VHS (Channels 7–13), and UHF (Channels 14–82). You will notice that the lower the channel number (lower the frequency), the greater the amount of distortion introduced into the picture. Lightning has a frequency distribution that is more concentrated at the lower frequencies than at the higher frequencies and transmission at lower frequencies is much more efficient at lower frequencies than at higher frequencies.

As the thunderstorm gets closer, you will notice that not only does the amount of distortion increase, but the number of visible distortion increases. The number of lightning "hits," the closer the storm or the greater the intensity of the storm. (Either way, take cover–it's about to hit!)

Those of you with an antenna rotator can actually determine the azimuth (location relative to North) of a thunderstorm to where the picture distortions are maximized in number and intensity. When distortions are maximized, you have determined the direction to or from the storm! If your antenna is directional, try orienting your antenna at an azimuth 180 degrees from where the distortion peaks. Whichever is greater (0 degrees or 180 degrees) is the direction from you to the storm.

Those of us in The Republic of Texas, have year-round practice doing this. Those of you in Maine may need to wait a few months to get your first chance to try it.

What does all this nonsense about lightning prove?

1) Amplitude modulated carriers (like TV "visual" carriers) are much more susceptible to noise intrusion than frequency modulated carriers (like the "aural" carrier).

2) The lower the frequency, the more efficient the transmission of signals, whether noise (like lightning strikes) or signal. It's harder to push higher-frequency signals through a cable than lower-frequency signals.
3) Benjamin Franklin was right-on with his infamous kite experiment.

How do you minimize the distortions associated with the modulation process? First, you don't overmodulate. Whenever you overmodulate the carriers, you create all manner of strange signal artifacts. To keep from overmodulating the carrier, you normally monitor the applied baseband signal with a waveform monitor (for video) or a VU meter (for audio). Yeah, when they say a maximum of 100 IRE of video (with 40 IRE of sync), they mean it!!! Imagine what would happen to adjacent TV channels if you overmodulate the visual carrier!

Q: What is MATV?

A: Master Antenna TeleVision. A "broadband" communications system that simultaneously distributes several channels of television (visual and aural) information.

Q: What is CATV?

A: CATV (Community Antenna TeleVision) system does exactly the same thing as an MATV system EXCEPT it crosses building property lines. The CATV industry is regulated by the FCC (Federal Communications Commission) to perform to applicable paragraphs of Parts 73 and 76 of the FCC Rules and Regulations. Copies of those specifications are available from your local Government Printing Office (GPO) Bookstore or from the Superintendent of Documents in Washington, D.C. You can also find a copy of those documents in libraries that are Federal Depositories of Information.

Q: ***We use cable TV to distribute internal communications programs to 130 TV receivers around our manufacturing site. We get vertical jitters in the picture even after time base correction and ridding the signal of time code. We suspect that the video switcher's clamping circuit is being fooled by something and is causing the jitter. Any solutions? —James Wilkinson-Ray, IBM, Essex Junction, VT***

A: It sounds like you have eliminated most of the potential problems with the "baseband" (unmodulated) portion of the picture. You don't mention it, but I assume that the playback picture is stable before it goes to the modulator. This means that the problem is induced in one of the following:

1. The modulator (very low-power TV transmitter that takes the "baseband" video and audio signals and impresses them onto a TV channel);
2. The RF distribution system (cables, connectors, amplifiers, etc.);
3. The demodulator (tuner) circuits in the various receivers (that derive representations of the original "baseband" video and audio signals from the TV channel);
4. The deflection circuits in the various receivers.

It is highly improbable that several different receivers will exhibit exactly the same problem in exactly the same way (although I have seen some cases where the design was so faulty that the problem was induced in several sets of the same make and model). For this reason, we can usually discount the demodulator and deflection circuits in the various receivers (items 3 and 4 on our list).

The next step to finding the answer can be derived from the jittery picture. When a picture is jittery, the time at which the receivers start their vertical scans is not the same as the time at which the incoming signal says to start. Somewhere through the process, a problem is being induced in the interpretation of the vertical sync pulses in the signal. There are two ways in which this can happen:

1. The sync pulses in the signal are being suppressed (reduced in amplitude) to the point where the receivers cannot lock onto them
2. There is interference coming from the sync pulses in an interfering video signal

The suppression of the sync pulses (both vertical and horizontal) is slightly more likely than an interfering signal. In a modulator, when the video signal is impressed on the visual carrier, some modulators have trouble achieving "linear" operation across the entire voltage range of the video signal. This means that the picture portion (above "0" on the waveform monitor) may not get distorted, while the sync portion (below "0" on the waveform monitor) does get distorted. This problem is frequently induced by incorrect adjustment of the visual carrier modulation (usually excessive modulation). To adjust visual modulation, send 100% saturated color bars into the video input of the modulator and adjust for a 75% depth of modulation (as shown on the modulator meter or indicated on a spectrum analyzer). If you have neither, look at the demodulated video signal with a waveform monitor and adjust until there is no compression of the sync pulses.

The second potential cause of the problem–an interfering signal may be induced in the system by improper shielding. Any signal that has a frequency close to 59.94 Hz (including strong 60 Hz pulses from manufacturing equipment, and off-the-air signals that occupy the same TV channel as you are using) can create the jitter. It can get in through the cable, through microscopic gaps in connectors, through plastic receiver cases, or via any one of a number of other potential paths.

Take a demodulator (or a TV receiver with a video output) and a waveform monitor or oscilloscope to the point where one of the jittery receivers is located. Carefully examine the vertical blanking interval (at 2V MAG on the waveform monitor) for any interfering signal or sync compression.

Finding and correcting the problems induced by an interfering signal sometimes require the combined efforts of Miss Marple, Hercule Poirot, and the violin master himself. First, if you suspect an off-the-air interfering signal, wait until they go off the air to see if it goes away. If you suspect other sources of interference, you must eliminate them one-by-one. Look specifically at high-current operations (motors, wave soldering machines, etc.) and at motor control, light dimmer circuits and gas discharge lighting ballasts.

Once you find the source, it can be extremely difficult and costly to correct the problem. You may try changing the azimuth of the sets (assuming that the interference is localized and directional). You may try double-shielded cable (assuming that's where your weak spot is). You may try receivers with metal cases (assuming that the extra shielding will stop some of the interfering signal). You may try power line filters on the receivers and interfering equipment (assuming that the noise is going through the power line). You may try to replace the hair that you pull out trying to fix the problem (assuming that's replaceable).

Part II

Signals

General Signal Issues

Q: What is a dB?

A: The term dB stands for "deciBel" and expresses a ratio between two signals. The signal parameter that is being described may be current or voltage or power.

The ratio is logarithmic. A doubling of, say, signal voltage does *not* produce a 2dB gain. The formulas for determining dB ratios are:

$$\text{dB (power)} = 10 \times \log \left(\frac{\text{Power of Signal 1}}{\text{Power of Signal 2}} \right)$$

$$\text{dB (voltage)} = 20 \times \log \left(\frac{\text{Voltage of Signal 1}}{\text{Voltage of Signal 2}} \right)$$

$$\text{dB (current)} = 20 \times \log \left(\frac{\text{Current of Signal 1}}{\text{Current of Signal 2}} \right)$$

What does all of this technical gobbledygoop mean? Let's take an example. Let's say your camera produces a signal with a 58dB signal-to-noise ratio. Remember that the output signal from the camera is 1.0 Volts from sync tips to peak white. This means that the average (rms) noise signal has a voltage of

$$58\text{dB} = 20 \times \log \left(\frac{1.0 \text{ Volts}}{\text{Noise}} \right)$$

$$\text{Noise} = 0.00126 \text{ Volts}$$

Q: What is the signal-to-noise ratio?

A: The signal-to-noise ratio is just what the title implies–it's the ratio, expressed in dB, between the peak voltage of the desired signal and the average (rms) voltage of the undesired noise. It's the expression of how much background noise is present from a loudspeaker or "snow" in a picture.

Be very careful when comparing the absolute signal-to-noise numbers. There are billions and billions of ways of measuring signal-to-noise ratio. On a camera, any disclaimer listed with the signal-to-noise ratio must be considered when comparing numbers. Typical disclaimers include weighted, bandpass limited, luminance only, chrominance only, and gain boost at "0."

Q: What is phase cancellation?

A: First, what is phase? Phase refers to the timing of the voltage peaks and valleys of an A.C. (*alternating* current) signal. As shown in Figure 12, the voltage of an A.C. signal varies at a rate ("frequency") between a "high" and a "low" voltage.

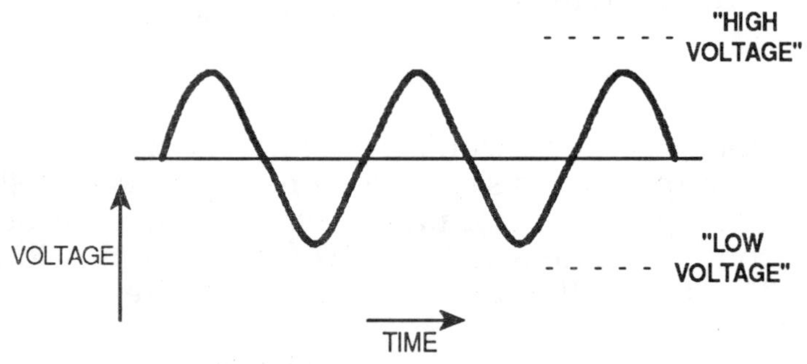

Figure 12
An AC Signal Varies Between a "High" and a "Low" Voltage

As shown in Figure 13, when adding two signals of the same frequency, if the peaks and valleys occur at the same time, you get double the peak-to-peak voltage (and double the fun). The two signals are said to be "in-phase."

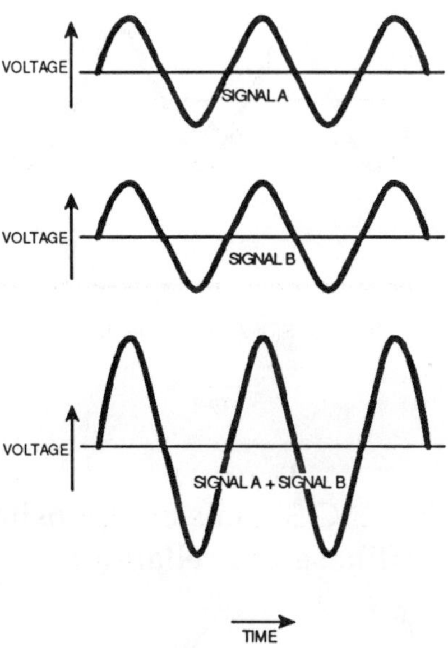

Figure 13
Adding Two AC Signals of the Same Phase

Figure 14 shows what happens when the two signals are exactly the opposite ("180 degrees out-of-phase"). At any given instant, the voltage from Signal A is exactly equal to the negative voltage from Signal B. When Signal A is added to Signal B, the net result is zero (zip, cero, nil, nought).

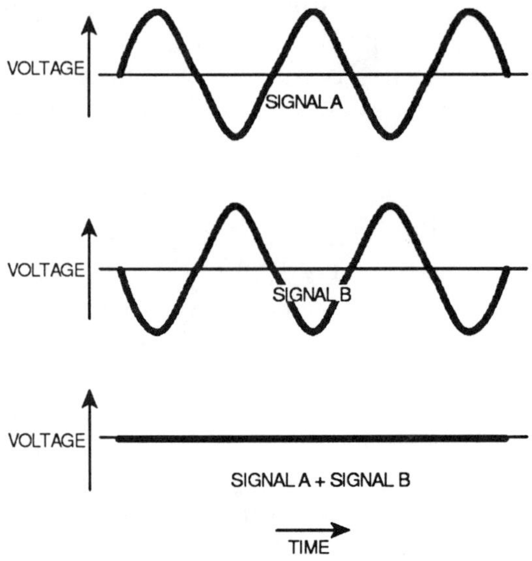

Figure 14
Adding Two AC Signals of Opposite Phase
(Phase Cancellation)

When must we worry about such things? Always. We run into the phase cancellation problem in acoustics, audio, video, data, and radio frequency (RF) signals.

Let's take audio as an example. Suppose we have two microphones connected together and are picking up the same sound source, as shown in Figure 15. The microphones are equidistant from the sound source. The sound pressure waves (varying between "high" and "low" pressure) arrive at both microphones at the same time. If the wires connected to the microphones are connected so that each microphone creates a "high" voltage in response to a "high" pressure, no problem; however, if you reverse the wires connecting *one* of the microphones, we got a problem. The signal from the microphone on which you reversed the wires will create a "low" signal voltage in response to a "high" pressure wave. Microphone A creates a "high" voltage at exactly the same times that Microphone B creates a "low" voltage. Add the two signals together and whaddayouget? The dreaded phase cancellation result of nothing.

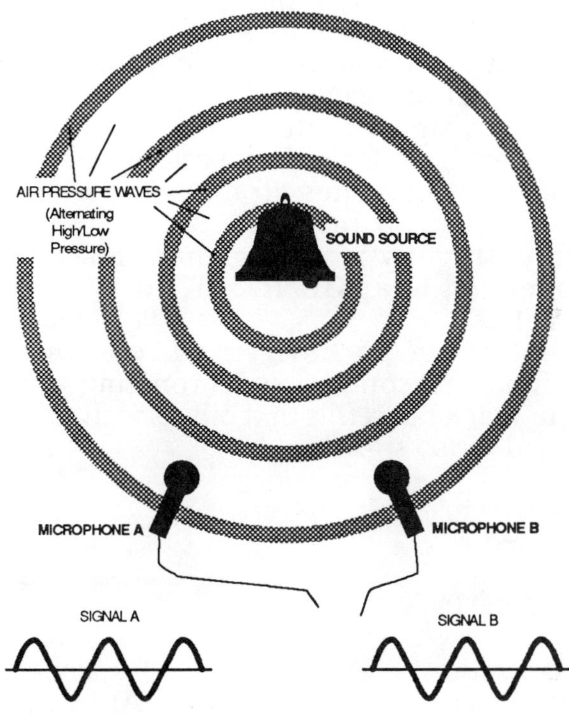

Figure 15
Phase Cancellation of Microphone Signals

If you stop to think about it, in many cases, one microphone will produce a greater signal than adding the signals from two microphones. Neat, huh? More for less! THAT ought to keep your manager happy!

Q: What is the difference between "Sync" and "Black Burst?"

A: Sync is a signal composed of pulses (at 4 Volts peak-to-peak) that signifies the start times of each horizontal and vertical scan in the picture. Sync contains no information that can be used to lock the color operations of a piece of equipment. Sync is connected in a "sync lock" system to make equipment scan synchronously. Sync is normally connected to a "SYNC" input connector.

Black Burst is a signal composed of sync (at 0.286 Volts), black picture level, and color burst reference signal (at 0.286 Volts). Black Burst contains all the information needed to "genlock" a system to make the scanning and the color generation to properly work together. Black Burst is frequently connected to a "GENLOCK" or a "REFERENCE" input connector.

Remember that a genlock system will operate by referencing to any signal that contains synchronizing and color burst information. Black Burst fills that bill, as well as color bar and other encoded video signals.

Q: What is HDTV?

A: High Definition TeleVision (HDTV) is probably going to be the next generation of television. Although there are ongoing discussions about HDTV, Advanced NTSC and other high-quality television systems, the Japanese-sponsored HDTV system is out ahead in delivering equipment.

The Japanese version of HDTV uses 1125 horizontal scans (instead of our normal 525) that are refreshed every second (the same as our current standard). HDTV at present has no encoding standard that combines black-and-white and color information onto one cable. There are, however, many proposed bandwidth conservation systems.

This whole mess is all caught-up in politics. The US officially adopted the Japanese system a few years back. Now American manufacturers are fighting back with their own proposed standard–and the FCC is listening. The Japanese system has been pummelled by the Europeans for not being compatible with their existing standard and by the US for not being compatible with their standard.

When is this going to happen? Who knows. Depending on when the standards discussions start to settle down, we'll probably see the initial rush of HDTV in the USA in about three or four years. There are several "ifs ands or buts" involved here, but to me, this timetable appears reasonable.

Q: What do you think about HDTV?

A: Here's my opinion:

I hate to say it, but I think "HDTV" ("High Definition TeleVision") probably will not reach the consumer in the near future. There will be a market for "HDTV" (in its current form), but I don't think you will see it in your living room in the next five years.

Yes, "HDTV" displays a picture that is markedly superior to existing NTSC pictures. Personally, I've been watching "HDTV" in its current form for almost a decade. For someone who considers bars-and-tone an erotic experience, "HDTV" makes for considerable embarrassment.

I don't think that the picture improvement is enough motivation for you to dig $3,000 out of your pocket for a new, improved, incompatible receiver. (Particularly when consumer loan defaults are the highest they've been in years.) I also do not think that the picture quality is enough for each television station that you watch to justify $25-30,000,000 on "improvements" to obsolete their existing investment. It simply does not make economic sense.

Significant improvements are coming our way with IDTV (Improved Definition TeleVision) and EDTV (Enhanced Definition TeleVision). These are systems that are considerably lower in cost, compatible with existing receivers and broadcast facilities, and present a display that is a quantum improvement over what you now watch. When? Could be as early as this year when we begin to see the results.

What will happen to the latest state-of-the-art technology developed for HDTV? Will all that development go to waste?

To those questions, I must ask WHAT state-of-the-art technology? The proposed HDTV standard is predicated on the same tube-type technology that limited NTSC signals *half a century ago.* The proposed HDTV cameras are based on the same imaging technology you've had in your hands for decades. HDTV imaging still uses the same display tubes and projection technology that was available when Big Band sounds were popular. HDTV transmission, with its necessary bandwidth reduction schemes, will severely reduce the quality of the pictures that can be delivered to your home. (Indeed, there is no existing transmission structure in North America that can be readily and economically adapted to HDTV use.) Don't listen to the smoke being blown by ill-informed bureaucrats, politicians, and marketeers.

There will be some applications that can benefit from the HDTV effort– electronic cinema theatres, scientific research and monitoring, industrial research and monitoring. There is no infrastructure currently in place that has the capability to get an HDTV signal from production to the home viewer. It's improbable that you will have a wide-screen high-quality picture in your living room in the near future. What's it going to buy you–the ability to watch "Roseanne" with a wide aspect ratio?

Let's put the hype to rest and get on with our business. Let's return to the basics of educating, communicating, and entertaining–**we've ignored them long enough**.

Q: Is there any way to measure the amount of delay between two video signals?

A: Sure. Lock up a waveform monitor to external sync. Switch the display for "1 μs/div" (one microsecond-per-division). Look at one of the signals to be measured. Vertically center the sync pulses on that signal so that half the voltage (about 20IRE) is above the hash-marked graticule line and half the voltage is below. Horizontally adjust this sync pulse so that the left edge (the descending voltage side) of the sync pulse is on a convenient hash mark. Note which hash mark you are using.

Without changing the external sync source or the horizontal display adjustment, display the other signal to be measured. Vertically adjust the display so that half the sync voltage is above the hash-marked graticule line and half is below. Count the number of hash marks separate the sync signal from the second display from the first. (Each wider hash mark constitutes one division. The shorter hash marks denote fractions of one division.)

The number of divisions represents the number of microseconds of delay between the two signals. Depending on the magnitude of delay and which signal is delayed, you may need to repeat the process using another hash-mark reference.

This procedure measures the difference in H-Phase between two signals. Similar measurements can be made to determine the amount of delay between any two picture elements.

Q: How can I simulate a telephone conversation for my audio?

A: There are three basic factors–bandwidth, presence and noise–that make a telephone sound as bad as it does.

Bandwidth can be simulated by getting a bandpass filter that only passes those portions of an audio signal that have a frequency between 200Hz and 3.7kHz. This doesn't have to be too exact, but the closer the better. Such a filter is available from several manufacturers including Sescom, or you can roll yer own by building a system with parts from Radio Shack (or others).

The second factor is noise. (It seems ironic that telephony engineers have worked for decades to reduce this factor, but we have to put it back in.) A white noise generator would do the trick, provided you pumped this noise through the bandpass filter. If you don't have one, try the audio out from a TV set tuned to an unused channel. Sometimes you may need the noise, sometimes not. Play around to see what works for your application.

The third parameter is presence. When you speak into a telephone, your lips are only about an inch from the microphone. Do the same with your microphone. This is one time when a cheap microphone works best.

Q: I picked up a hum in my audio during production. Can I fix it in post?

A: Maybe. Most hums come at frequencies of 60Hz and 120Hz. If you have or design a band stop filter for those particular frequencies, you may be able to get rid of some of the interference. The problem here is that there is often desirable information at those frequencies that will also be filtered out.

There's another option that sometimes works. Generate a signal with a frequency of 60Hz or 120Hz. Add this signal with the audio signal and adjust the phase of the generated signal to minimum hum in the audio. Then adjust the level of the generated signal to minimize buss in the audio. Under some conditions, this may almost eliminate the interfering signal without distorting the desired signal.

The phase cancellation technique is touchy and requires rather sophisticated equipment, but when the material warrants it, there is no substitute.

Q: How can I calculate the exact amount of video signal delay that will be introduced by a cable?

A: (This one, of necessity, takes some math. It will help to use the old engineering school trick of "cancelling the units", where $\frac{\text{seconds}}{\text{seconds}}$ gives no units of measurement.)

The delay from a given type of cable is dependent on the "velocity of propagation" from that piece of cable. You can find this number in the specifications for the cable. For instance, for Belden 8281 precision video cable, the nominal velocity of propagation is 66%. Some 75-Ohm cables have a nominal velocity of propagation of 66%, others have 78%. In order to calculate the delay you must know the velocity of propagation of your cable.

For the sake of argument, let's assume that your cable has a velocity of propagation of 66%. This number means that the signal travels through that cable at a speed equivalent to 66% of the speed of light through free space. The speed of light through free space is 186,282.429 miles-per-second (close to the now-famous 186,000 miles-per-second). So, with your handy-dandy calculator (preferably the one between your ears), this means that the speed of the signal down this cable is 122,946.40 miles-per-second which is equal to 7,789,884,107 inches-per-second. Whew.

For a six foot length of cable (equaling 72 inches), it takes

$$72 \text{ inches} \times \frac{1 \text{ second}}{7789884107 \text{ inches}} = 0.00000000924 \text{ seconds}$$

to go down that cable. Remembering our basic units of measurements:

$$0.00000000924 \text{ seconds} = 9.24 \text{ nanoseconds}$$

Yes Mork, nanoseconds. This is the amount of delay that would be encountered in six feet of Belden 8281 cable.

In terms of degrees of change in the NTSC subcarrier signal at a frequency near 3.58 MHz, this six-inch cable delay equates to

$$0.00000000924 \text{ seconds} \times \frac{3579545.45 \text{ cycles}}{1 \text{ second}} = 0.033 \text{ cycle}$$

$$0.033 \text{ subcarrier cycle} \times \frac{360 \text{ degrees}}{1 \text{ cycle}} = 11.91 \text{ degrees}$$

Now we know that a six-foot piece of cable will introduce 9.24 nanoseconds = 11.91 degrees of subcarrier delay.

Can this introduce visible distortions in the picture? Youbetcha. If you switch between two signals with a chrominance phase difference greater than about five degrees, you can see the hue shift in the picture.

Can you see the picture shift position on a picture monitor?

Probably not. Assume you have a 19" (nominal diagonal measure) picture monitor. From ye olde Pythagorean theorem, this means that the width of the picture is about 15.2". We know (from earlier outstanding articles) that the active picture (0.0000635 seconds total minus 0.0000114 of blanking) takes up about 0.0000521 seconds of time during each horizontal scan. So, with a little math (ugh) we can calculate the rate at which the picture monitor must write the scanned picture.

$$\frac{\text{15.2 inches}}{\text{1 horizontal scan}} \times \frac{\text{1 horizontal scan}}{\text{0.0000521 seconds}} = 291746.64\ \frac{\text{inches}}{\text{second}}$$

A six-inch piece of Belden 8281 cable will shift the picture by

$$0.00000000924 \text{ seconds} \times \frac{\text{291746.64 inches}}{\text{1 second}} = 0.0027 \text{ inches}$$

on a 19" monitor. (All this stuff assumes that the size of the available scanned picture is exactly the same size as the mask of the monitor in front of the picture tube. It might be an interesting exercise to calculate the size of the shift with the nominal 15% of picture overscan found in most picture monitors. Hint–the answer is 0.0031 inches.)

I'm tired.

At no extra charge, here's a little BASIC computer program to calculate cable delay (you may have to make some syntax changes for your particular version of BASIC):

```
10 INPUT "VELOCITY OF PROPAGATION (%)";V
20 INPUT "CABLE LENGTH (INCHES)";L
30 C=11802854700
40 S=.01*V*C
50 D=L/S
55 Z=D*10^-9
60 H=1288636362*D
70 PRINT "DELAY ENCOUNTERED IN ";L;" INCHES OF
     CABLE WITH A ;V;"% VELOCITY OF PROPAGA-
     TION ="
80 PRINT "     ";D;" SECONDS = ";Z;" NANOSECONDS"
90 PRINT "     ";H;" DEGREES OF 3.58MHz DELAY"
100 END
```

Fractured Definition:

blank·ing \ *'bla ŋk-ing* \ v ***1a:*** act of removing selected expletives from a proposed script ***1b:*** state of censor when selected expletives were not removed from a proposed script ***2:*** act of selective neurocide performed with drugs or alcohol ***3:*** state of a computer screen immediately before a file is saved ***4:*** state of employment of a technician who ignores video signal standards

n ***1:*** the voltage level to which the video signal returns during picture retrace ***2:*** the time interval in a video signal allocated for picture retrace

Q: What is "super black?"

A: Super black is the adjustment of pedestal when developing special effects so that is below blanking. The relocation of pedestal from +7.5 IRE to about -20 IRE (as measured on a waveform monitor) may improve the quality of a key or of an inserted background matte.

The increased difference of levels between the desired foreground to be keyed and the super black background decreases the amount of time it takes for a key circuit to recognize the key transitions. This reduction of recognition time improves the sharpness of the keyed outline.

Super black also reduces the possibility of seeing a background in the finished picture when digital video effects are used to manipulate the keyed image. At times, black level differences are visible in the picture. (If you want to see some examples, playback a copy of "Star Wars" and look at some of the shots where there are several space ships involved in a battle. Turn up the brightness on the TV set and look for black areas that follow almost each of the space ships. Although ***Star Wars*** was shot on film, the idea about super black in video allows correction of the same type of effect.)

A video signal with a -20 IRE super black level *does not* meet normal video specifications and cannot be transmitted by a television station. (Remember that pedestal–the blackest black in the picture–must have a nominal value of +7.5 IRE to reduce the possibility of the picture losing its sync reference.) The use of the super black is limited only to the time when an effect is being developed; even then, many pieces of equipment (VTRs, processors, monitors, etc.) have a hard time handling the out-of-specification signal. On many pieces of equipment, a "black clip" circuit will reduce the magnitude of the super black signal (usually to 0 IRE or above). Somewhere, the pedestal must be re-established at +7.5 IRE before the signal is transmitted. A processing amplifier (or the processing amplifier circuit in a time base corrector) can usually handle that chore.

Q: Why is there a +7.5 IRE pedestal?

A: Remember that the television standards were developed before the transistor was a gleam in William Shotkley's eye. Everything electronic was accomplished by burning up material in a glass-enclosed vacuum and manipulating the excess electrons emitted by the burning material. With these "tubes," sending electrons through an imperfect vacuum and the inevitable variations in the quantity of electrons available for manipulation created a very unstable operating environment.

In those days, it was hard to develop a television set that was stable enough to reliably separate the sync portion of the video signal from the picture portion. Because of this difficulty, a 7.5 IRE (equal to 0.054 Volts) fudge factor was built into the video signal specifications to create a "no man's land" and make it easier to for tube circuits to positively identify the sync portion of the signal.

Keep in mind that the pedestal actually reduces the contrast ratio of the television system by 7.5%. We could actually create a better picture if we could eliminate the pedestal.

Flash forward to today. New solid-state equipment does not really need the +7.5 IRE pedestal. Last year the FCC considered a proposal to eliminate the pedestal in their specifications. The proposal was defeated because new program material would not be exactly the same as old program material and there would be problems in maintaining a consistent black picture brightness. Everybody would be getting up and down, up and down, up and down readjusting the brightness on their TV set. Heaven knows that the American public doesn't need exercise like that–that's the drudgery that remote controls were invented for!!

Q: How do I know that my television signals are technically correct?

A: The ultimate judge is your intended audience. If you communicate the ideas to them with no picture or sound impairment, then your signals are probably at least close to correct. All of this measurement stuff with waveform monitors, vectorscopes, oscilloscopes, frequency counters, meters, and spectrum analyzers are to maintain a consistent level of quality of signal. If you're consistently happy with the final picture and sound presented to your intended viewers, then forget about the technical standards–the system is already working to your specification. On the other hand, if you occasionally have "glitches" in the picture or sound, pay attention to the basic measurements.

The most important pieces of measurement equipment are those "baby blues," "Clark Gable ears," and neurons that are strategically installed above your shoulders. You've got to be able to sense that there is a problem and develop solutions for correction of the problem.

The second-most important genre of test equipment are the devices that will be used to present your message to your audience. Normally these are picture monitors and sound systems that will hide an awful lot of distortion before becoming objectionable to the audience.

The third tier of test equipment is the expensive "engineering monitors," waveform monitors, vectorscopes, etc. This level of testing is designed to show you more of the distortions in the signal that will be evident on the type of equipment used for presentation to the audience. (If you want to see an absolutely horrible picture, connect an engineering-quality tuner to an engineering-quality picture monitor and watch for a little while. If you can, set a standard consumer TV set side-by-side. I'll bet you like the picture quality from the $300 TV set a lot better than the $15,000 engineering test set.)

The engineering equipment is designed to show you the least little error so that it can be corrected before it reaches a magnitude that is objectionable on the "lower quality" "consumer" types of equipment. If it looks good on the "high quality" "engineering" equipment, it is bound to look good on "consumer" equipment.

With "engineering" quality equipment, what do you look for? On an "engineering" picture monitor, look for any picture instability, look at the quality of the colors, and look for any ghosting of picture details. On a waveform monitor, make sure that the filter is switched to "IRE" (or "Luma" or "Low Pass") and make sure that the sync pulses are squared-off and go between blanking at "0 IRE" and "-40 IRE." Make sure that the picture portion of the signal doesn't go below "+7.5 IRE" or above "100 IRE." On a vectorscope that is looking at color bars, make sure that the dots representing the various colors fall in the large boxes.

It's that simple. Most operators don't need to know about differential gain, or chrominance-to-luminance delay, or transient intermodulation distortion, or any other such nonsensical tests. Those tests are there for technicians to check the technical quality of operation of the equipment–not for the operators of that equipment to make on-the-air adjustment.

There is only one time when there is a higher authority that will stand in judgement of the technical quality of your signals. That time only occurs when you throw the signal through the "ether" to the home television sets. In only that case, the Federal Communications Commission (FCC) will measure the technical quality of signals as they're broadcast and, if they're not up to snuff, they will let the broadcaster know.

If you're sitting on top of a corporate television facility that has no intentions of broadcasting the signal–you do not absolutely, positively have to measure up to all the standards. As long as you are happy with your results, don't try to "fix it." If you are not happy with the results, then you need all this other technical nonsense to try to figure out how to "fix it."

Q: Why doesn't television use R/G/B video signal components instead of NTSC, Y/C or Y/R-Y/B-Y signals?

A: First comes the consideration about the number of cables or channels required to convey all the information it takes to create a color picture. The NTSC (Never Twice the Same Color) signal squeezes the R/G/B (Red/Green/Blue video signals) onto a single cable. When you use a single cable, the video equipment requires only one channel through which the signal must be switched, processed, or transmitted–it's "inexpensive."

Y/C (luminance/chrominance) separates the two major portions of the NTSC signal and processes them separately. The Y ("luminance"–picture "black-and-white") portion of the signal is a particular combination of the R/G/B signals. For all analog television systems (including NTSC PAL and SECAM), the proportions for Y are as follows:

Y = 30% of the Red + 59% of the Green + 11% of the Blue.

The C ("chrominance"–picture "color") portion of the signal is the result of combining two other combinations of the R/G/B signals that have been amplitude and phase modulated onto 3.58MHz (in North America) "carriers." When used in this sense, a carrier is a modulated signal (similar to a radio frequency -"RF"- signal). Actually these chrominance signals are called "subcarriers" because, when transmitted, they convey the color picture information within another "visual" RF carrier that carries both luminance as well as chrominance.

The two chrominance subcarriers in North America are

called "I" and "Q." (See the next question about other television systems.) In the simplest form, I+Q = C. The formulas for the proportions of R/G/B information within these two subcarriers are as follows:

I = 60% of the Red - 28% of the Green - 32% of the Blue

Q= 21% of the Red - 52% of the Green + 31% of the Blue

One thing to keep in mind here, Y+C = NTSC. In NTSC (single channel) systems, the Y and the C actually interfere with each other, creating picture distortions that are objectionable in some applications.

Y/C systems require two channels–slightly greater expense to gain an extra generation of video tape capability. By keeping the luminance information separated from its associated chrominance, they cannot interfere with each other.

Y/R-Y/B-Y (luminance/Red minus luminance/Blue minus luminance) "component" systems require three channels for proper operation. Here, the Y/R-Y/B-Y components are NOT modulated onto subcarriers; instead, they are used in their original "baseband" form. These components are derived as follows:

Y = 30% Red + 59% Green + 11% Blue

R-Y = 100% Red - (30% Red + 59% Green + 11% Blue)
= 70% Red - 59% Green - 11% Blue

B-Y = 100% Blue - (30% Red + 59% Green + 11% Blue)
= -30% Red -59% Green + 89% Blue

As you can see from the formulas, the Red, Green, and Blue signals are closely interrelated among the three combined signals. If you distort any of the individual combined signals, you will see the difference, but it would not be as great a difference as if the system were purely R/G/B.

There are other problems with noise and required signal bandwidth associated with R/G/B component signals. In short, with existing technology, R/G/B simply is not as good a signal to record, process, or transmit as the combined signals as the combinational signals.

The basic formulas can be summarized as follows:

NTSC = Y + C

where,

Y = Unmodulated picture black-and-white brightness "base"

C = Modulated picture color information
= Modulated I Subcarrier + Modulated Q Subcarrier

R-Y = Unmodulated combination "component"

B-Y = Unmodulated combination "component"

Q: What are I and Q?

A: The answer isn't pretty for this one. Basically, I and Q are components of chrominance (C). Chrominance is the portion of the video signal that describes the color that will appear in the picture.

I + Q = Chrominance

When you talk about chrominance, you are talking about the combined I and Q signals. I, Q, and chrominance all have an at-rest frequency of 3,579,545.45 cycles-per-second (3.58 MHz).

I has a phase delay of 33 degrees relative to system subcarrier. Q has a phase delay of 123-degrees relative to system subcarrier. Notice that 123-33 means that the phase difference between the base 3.58MHz signals is 90 degrees.

The signal used to modulate the I subcarrier is composed of 60% of the video signal from the red tube or chip minus 28% of the video signal from the green tube or chip minus 32% of the video signal from the blue tube or chip. We can abbreviate this as:

I Modulating Signal = 0.6R -0.28G -0.32B

The Q component is modulated with a signal consisting of:

Q Modulating Signal = 0.21R -0.52G + 0.31B

Now we have the two basic differences between I and Q: the phase of the subcarrier signal is different (by 90 degrees) and the modulating signals convey different combinations of the red, green, and blue video signals.

It requires both the I and Q signals to describe color. The resulting phase of the combined I and Q signals is compared to the color burst (at the beginning of each horizontal scan) to convey color hue information. The resulting instantaneous amplitude of the combined I and Q signals convey color saturation information.

Just remember that I and Q combine to give the chrominance component (C) of an NTSC encoded video signal.

Q: What would happen if you displayed an NTSC-M tape on a PAL-B/G color picture monitor?

A: If you go to Frankfurt, you can see this if you check into a hotel and watch the Armed Forces Radio and Television System station.

The picture will be black-and-white. The European-standard monitor is looking for chrominance (picture color information) at a frequency of 4.43361875 MHz. The chrominance signal from the station is conveyed at a frequency of 3.579545 MHz. This is like looking at Channel 4 for the picture being transmitted on Channel 5.

The other problem with the picture is that the monitor is designed to scan 625 horizontal scans in each frame. The transmitted signal has 525 horizontal scans in each frame. Although some monitors can't handle it at all, if the monitor works, the missing 100 horizontal scans will appear as black (unscanned) areas above and below the picture.

Sound may also be a problem. Most European receivers look for sound (on an "aural carrier") at a frequency that is 5.5 MHz above the picture (on a "visual carrier"). Normal North American transmission occurs with a 4.5 MHz separation between visual and aural carriers. Some receivers can handle the difference, others cannot. (Keep in mind that this discussion is about radio frequency–"R.F."–signals, not the baseband signals from microphones and mixers. At baseband, European and North American systems are directly compatible.)

... Regarding the display of NTSC on a PAL monitor: You are quite right about the problems of the monitor failing to "see" the 3.579545-MHz NTSC subcarrier because its chroma circuits are tuned to the 4.43361875 PAL chroma frequency. However, I have a problem with your closing paragraph, " this discussion is about (RF) signals... At baseband, European and North American systems are directly compatible. – Mr. Robert Hurst of the General Electric Government Services Division

What we have here is a failure to communicate. Halfway through my answer, I switched from video to audio. As Mr. Hurst observes, North American and European baseband video systems are incompatible; however, I am right because North American and European based audio systems are compatible. The confusion comes from my lousy writing style coupled with an aberrant mind. (There may be level inequalities, but an audio signal is an audio signal is an audio signal.) I wish to state unequivocally–I am always right, except when I am wrong.

Q: What are the frequencies associated with NTSC television?

A: There are several misconceptions about some of the frequencies associated with NTSC-M color television (the legal standard in all of North America).

First, the horizontal scanning frequency of color signals is 15,734.264 Hertz, *not* 15,750 Hertz as has been repeatedly (and erroneously) touted. The old number of 15,750 is for black-and-white signals *only*.Second, the vertical scanning frequency of NTSC-M color signals is 59.94 Hertz, *not* 60 Hertz. Black-and-white signals are the only television signals that have a 60 Hertz rate. Again, there is a lot of bad information that has been published out there.

Most sources quote the frequency at which chrominance (color) information is conveyed usually truncate the real number to be 3.58 MegaHertz. The actual number is really

$$\frac{63}{88} \times 5 \text{ MegaHertz} = 3.5479545.455\ldots \text{ MegaHertz.}$$

Be careful as you study these numbers, there is a lot of bad information about this that has been published. Let's try to get it right, folks. This becomes more and more important as we start trying to interface digital computers with the old analog television system.

Q: What do "Y/C 3.58," "Y/C 629," "Y-688," and "4:2:2" mean? –Robbie Chung, TimeWise Multi-Media Inc., San Francisco

A: What we need is an alphabet with more than twenty-six letters and a mathematical system with more than ten different-looking numbers. That oughtta keep the marketeers happy! I just hope that they don't discover the extended character set on computer keyboards! Here's definitions for some of the more common abbreviations that are being bantered about in the television industry:

Y = luminance (black-and-white overall picture brightness information)

C = chrominance (color picture information)

Y/C = luminance and chrominance of the same signal are simultaneously sent down two different cables (the two cables may be in the same cable sheath)

EIEIO = proposed international standard for farm video (originated at Old MacDonald's place)

3.58 = "Normal" NTSC chrominance that is conveyed at 3,579,545,4545455 cycles-per-second (about 3.58 million cycles-per-second)

629 = NTSC chrominance that has been heterodyned to a frequency of 629,000 cycles-per-second as found in VHS and S-VHS recording systems

688 = NTSC chrominance that has been heterodyned to a frequency of 688,000 cycles-per-second as found in BetaMax®, ¾" (U-Matic®), and ¾"-SP® recording systems

743 = NTSC chrominance that has been heterodyned to a frequency of 743,000 cycles-per-second as found in 8mm and Hi-8® recording systems

4:2:2 = Digital video sampling rate of "Luminance: Chrominance Axis #1: Chrominance Axis #2" (expressed as an approximate multiple of the chrominance subcarrier frequency)–used in D-1 digital recording systems

4:0:0 = Digital video sampling rates used in D-2 digital recording systems

Q: NTSC has a bandwidth of 4.2 MHz but the horizontal scans at 15.734 kHz. If I am doing my calculation right, then we can only have 260 dots-per-horizontal scan. So why do we have 700-line cameras? –Robbie Chung again!

A: I'd approach this calculation of the number of dots-per-horizontal scan a little differently.

First, we know that the duration of a horizontal scan is

$$\frac{1 \text{ second}}{15{,}734 \text{ horizontal scans}} = 0.0000635 \text{ seconds} = 63.5\,\mu\text{s (microseconds)}$$

Of that 63.5 μs, approximately 11.5μs of blanking is left out-of-the-picture (pun intended), leaving about 52 μs of each scan that is available for picture information. In each 52 μs interval, within a 4.2 MHz bandwidth (the FCC-specified bandwidth for the visual carrier is actually closer to 4.18 MHz), you can have up to

$$\frac{4{,}200{,}000 \text{ cycles}}{1 \text{ second}} \times 0.000052 \text{ seconds} = 218 \text{ cycles}$$

in each horizontal scan. Remember that a cycle consists of a peak and a valley of voltage, creating a white dot (from the voltage peak) and an adjacent black dot (from the voltage valley). Without considering any other factors, there are actually

$$218 \times 2 = 436.8 \text{ dots}$$

on each horizontal scan. This might lead you to believe that the picture has a limiting horizontal resolution of 436.8 lines–ZZZZT! WRONG!

In television resolution, we talk about pairs of dots, but there is another factor in expressing resolution. Aha–the oregano of television resolution! All measurements of resolution in television, including horizontal resolution, are made with respect to picture height. Huh?

This means that we need to introduce a factor of ¾ into our calculations. The ¾ comes from the 4:3 aspect ratio of the television screen. Now our 436.8 lines gets downsized to 327.6 lines of television resolution.

As a general rule of thumb, approximately 80 TVL (TV Lines) of horizontal resolution will be passed through a system for each 1 MHz of video signal bandwidth.

That discussion talks about the number of individual dots that may be seen. Now let's talk about the quality of the dots that you can see in a picture.

Imagine a single teeny white dot on a black television screen. Think about the edges of the dot. Does it look like somebody evenly sprayed white paint through a stencil onto the screen? Does it look fuzzy, like somebody sprayed paint without a stencil? The quality of the edges of the dots, as well as the number of dots is an indication of the sharpness of the picture.

As this dot gets processed through the system, it will naturally tend to become increasingly fuzzy. Noise and processing increases the amount of fuzz, slowly eating away toward the center of the dot. Eventually the dot starts looking like a hair ball.

If you start out with a nice clean dot, it can withstand a lot more fuzziation than a dot that is fuzzy to begin with. (Don't you love fuzziation?? I can see that as a basis for a 900-number service!)

When you start with a 700-line camera, the dots are pretty clean and stencil-sharp. When you start with a 400-line camera, the dots are already fuzzy. When you send these dots through a 300-line recorder, they both encounter the same amount of fuzziation. It's just that the amount of fuzziation for the existing fuzz on the dots from a 400-line camera is relatively greater than the existing fuzz on the dots from a 700-line camera.

That's why you want to start with the highest possible quality camera signal even though it may be significantly degraded later. It has to do with the quality as well as the quantity of the lines of resolution.

Fuzzy Wuzzy was a bear...

Q: What's oregano?

A: It's a spice sometimes found in spaghetti sauce. Watch ***The Andy Griffith Show*** and ***Andy of Mayberry*** reruns. You'll eventually find out why I call it the secret ingredient. I'm easily amused.

BITS & PIECES

I was thumbing through some papers the other day and found the following design elements detailed in the EIA RS-189A color bar specifications:

"(a) The positive peak levels of the yellow and cyan bars are nominally equal to reference white.

(b) The negative peak level of the green bar is nominally equal to reference black level.

(c) The negative peak levels of the red and blue bars are nominally equal."

Maybe that little bit of information will help you with color bar setup.

Q: Why is Y used to signify luminance?

A: I'm not sure. I spent most of Memorial Day looking for an answer and here's what I found:

Y hasn't always been used to signify the luminance portion of the television signal. In the fifth (1960) through the seventh (1985) editions of the ***NAB Engineering Handbook,*** the same section appears calling the luminance channel "M" for "monochrome." In some of the CIE (Commission Internationale de l'Eclairage) discussions, I found that "L" was frequently used to signify luminance.

But then I found additional CIE literature talking about the axes used to describe colors. There are two sets of axes established by the CIE in a standard adopted in 1931. One set of axes used our friends Red (R), Green (G) and Blue (B) to describe colors. Then a second set of axes, closely related to the R, G and B axes, was discussed. You guessed it, these three axes were X, Y and Z–the "Y" axis contains all the luminance information needed to describe colors. It is possible to use vector algebra and matrix manipulation to convert between the RGB and the XYZ axes.

Much of the electronic television standards information does not deal with Y as meaning a luminance signal; instead, the accepted standard is E_Y. The "E" stands for electromotive force, i.e. voltage. The Red, Green and Blue video signals are abbreviated E_R, E_G, and E_B, respectively. In a like manner the two chrominance axes are abbreviated E_I and E_Q. Thus, the FCC formulae read like this:

$$E_Y' = 0.30\,E_R' + 0.59\,E_G' + 0.11\,E_B'$$
$$E_I' = -0.27\,(E_B' - E_Y') + 0.74\,(E_R' - E_Y')$$
$$E_Q' = 0.41\,(E_B' - E_Y') + 0.48\,(E_R' - E_Y')$$

(The prime notation "' " means that the signals have been gamma corrected.)

Don't let the math throw you on this one–an explanation follows. The E_Y' signal goes through the encoder basically unchanged. The chrominance portion (E_C) of the signal is derived from the following formula:

$$E_C = E_I' \sin(wt + 33°) + E_Q' \cos(wt + 33°)$$

The sin (sine) and cos (cosine) factors indicate that there is a phase difference that must be considered when adding the E_I' and E_Q' signals together. That's also where the "I" for "In phase" and the "Q" for "Quadrature phase" (90° difference) nomenclature comes from.

Whenever the marketeers start talking about Y, C, R-Y, and B-Y, you will probably see these same signals in technical literature as being E_Y, E_C, $E_{(R-Y)}$ and $E_{(B-Y)}$. It's just another way of saying the same thing for different audiences.

If you need more information, the best synopsis that I've found is in the **Television Engineering Handbook** edited by the late K. Blair Benson. Be prepared with your mathematical wading boots!

Q: Why are there 525 horizontal scans -per-frame in the North American TV scanning standard?

A: First, it wasn't always 525 horizontal scans -per-frame. There were all sorts of scan rates originally considered, including 60-400 (variable), 120, 225, 240, 343, 375, 440-445 (variable), 405, 430, 441, 495, 507, 525, 550, and 605. The original recommendation of the NTSC was for 441 horizontal scans -per-frame, then they reconsidered the recommendation after reviewing a paper by Donald G. Fink. All of these odd-ball numbers are derived from the tube-based technology limitations that were imposed in the 1930's. Even with those limitations, the committees came up with a standard that has withstood the test of time–only now, sixty years later, have we pushed technology to the full limits of those original specifications.

The first problem with the scanning standard was associated with the amount of video signal bandwidth (range of frequencies) that could be pumped through tube-type amplifiers in transmitters and receivers. At first, it was thought that 0-2.5 MHz was all that could ever be economically done. Then there was a revelation that there might, some day in the future, be a technology that would allow a wider bandwidth. It was then assumed that the bandwidth could be around 0-4.0 MHz. (The final number turned out to be 0-4.18 MHz–a number that we still live with as the bandwidth of the visual carrier in transmitted TV channels.) The increased range of frequencies meant that a video signal with more scan lines could be passed with sufficient resolution to create an acceptable picture.

So the game was afoot to figure out the best way to increase the quality of the picture to take advantage of future technological developments. The 411 horizontal scan rate was used as the basis for comparison with other rates. Some basic parameters had to be considered.

First, picture resolution along the vertical axis of the picture had to be pretty close to picture resolution along the horizontal axis of the picture. (Remember that the number of horizontal scan lines is the primary determining factor in vertical resolution.)

Second, tube technology imposed a practical limit on how the synchronizing signals could be economically manipulated within television receivers to create the effect of scanning.

Third, the number of tubes required to manipulate the synchronizing signals within any given receiver needed to be reduced.

Fourth, an odd number of horizontal scan paths were needed to achieve interlaced scanning–a way to more effectively use precious bandwidth.

Fink took a close look at the range of values between 400 and 600 and a pattern started to emerge. He came up with the following table of factors:

Number of Lines	Factors	Percentage of 441
405	3 × 3 × 3 × 3 × 5	92%
441	7 × 7 × 3 × 3	100%
495	3× 3 × 5 × 11	112%
507	3 × 13 × 13	115%
525	3 × 5 × 5 × 7	119%
567	3 × 3 × 3 × 3 × 7	129%

The "higher" 11 and 13 factors in the 495 and 507 rates were difficult to economically accomplish with the tube technology. The higher number of factors in the 405 and 567 lines rates were discounted because a higher number of circuits and tubes would be required to manipulate them. This left 441 and 525 line rates.

The 525 line rate increased vertical resolution by 19% over the 441 line rate while decreasing horizontal resolution by 16%. The 525 line rate took better advantage of foreseen advances in technology instead of saddling the system with existing technology limitations. Thank you, Mr. Fink.

In addition to the questions about line rates, there were also questions about frame rates. They finally standardized on 30 frames-per-second for black-and-white television. This became the standard because it was the lowest fraction of the well-established A.C. power frequency (60 Hz) that didn't produce objectionable flicker in the picture. It was important to have this power/television frequency interrelationship to minimize picture hum that was rampant when using tube technologies. This same rationale was also used to reject the notion of using the same 24 frames-per-second rate in motion picture films with sound.

Now, it appears that the latest technological developments can be incorporated into the standard without forcing you to throw away your existing television set (as the proposed HDTV standards would have you do). Let's hope that one of these Enhanced Definition TV (EDTV) or Improved Definition TV (IDTV) standards is adopted by the current testing being done in Alexandria, Virginia. I don't want to buy HDTV sets and place them next to my existing receivers–particularly at a couple of grand a pop!

If you really want to get your neural juices flowing, think about all the alternative scanning standards that are possible. Why scan at a constant rate? Why scan in one direction (left-to-right, top-to-bottom) only? Why not scan in a circle outward? Why scan at all–why not use the new technology to create bit-mapped planes directly from the imaging chips? As Bo Pilgrim would say, "It's a mind-bogglin' thing."

If you want to read more about the development of the television standards, dig around until you find a copy of **Television Standards and Practice: Selected Papers from the Proceedings of The National Television System Committee and Its Panels**, *edited by Donald G. Fink (New York: McGraw-Hill, 1943).*

Signal Standards

Q: What is RS-170A RGB?

A: Ain't no sech animal. RGB means Red, Green, and Blue primary color *component* signals. RS-170A refers to an NTSC encoded *composite* signal. Encoding takes RGB signals and converts them into an RS-170A signal. Don't let those computer guys fool you!

Wayne Dengel of Qutron writes to confirm many of my past comments about the computer genlocker market, citing additional problems with aspect ratio differences on various computer display standards and problems with resolving the flicker that often remains in the converted signal. Using today's technology and economics, the conversion from personal computers -to- television simply cannot be an inexpensive procedure. Yeah, you can get somewhat of a usable picture by creative processing. But, to the best of my knowledge, you cannot get broadcast-quality non-flickering pictures from anything but a frame store/scan converter.

In my opinion, this computer genlocker business is another one of those areas where there is more smoke than fire (like the "paperless office," "desktop publishing," and "inexpensive video bandwidth compression"). You cannot get the graphics of a $250,000 graphics package from a $10,000 box, regardless of what you may read or hear. The $10,000 box will neither have the picture quality nor the speed of generating the final product.

What you may get from an inexpensive system is acceptable pictures for your application. The only way you'll know is to go and look at 'em instead of reading about 'em. Look at the size of the discrete picture elements on a screen of the same size as used by your audience. Look carefully. Is there a particular "look" to the picture? Some computers create picture elements that are square, some rectangular. Some computers create low-resolution pictures with many colors, or high-resolution pictures with few colors. Some computers are so slow in operation that pen-and-ink is faster. These are limitations that cannot be overcome by simple conversion to television video.

If any particular manufacturer can't or won't demonstrate their product, be *very, very* suspicious. There are a lot of products in prototype configuration sitting in garages and basements. There are many complex technical issues in this computer-to-video conversion process that must be simultaneously solved.

Q: I have often seen statements that begin with "According to NTSC standards . . ." or "SMPTE . . .". How and where are these published? I would like to know where I can see these sources. —Jim Haba, 20/20 Video Elk Grove Village, IL

A: The NTSC was a committee formed by the Federal Communication Commission (FCC) to form a more perfect standard for encoding the red, green, and blue video signals into one all-encompassing video signal. The results from the NTSC action were incorporated into Part 73, Subpart E, Paragraphs 73.681 through 73.699 of the current FCC regulations. Some of the reading ain't easy! You can find these specifications in many libraries that are depositories of government publications or you can order from:

Superintendent of Documents
U.S. Government Printing Office
Washington, D.C. 20402

There's a fascinating history (I'm easily amused) of NTSC contained in the Federal Communications Commission Reports, Volume 41, Television Matters, September 1, 1950 to June 30, 1965. You can get this book from the Superintendent of Documents.

As far as SMPTE recommended practices are concerned, you can get information from:

Society of Motion Picture and Television Engineers
595 West Hartsdale Ave.
White Plains, NY 10607

You may also hear about EIA. EIA is an electronics manufacturer association that issues Recommended Standards (that's what the "RS" in RS-170A means). You can get information about their standards from:

Electronics Industries Association
2001 Eye St. NW
Washington, DC 20006

And yet a fourth popular standards group is the IEEE. These are the folks that brought you RS-232C and RS-422 for the control and communication of digital equipment. You can get copies of their standards from:

Institute of Electrical and Electronics Engineers
345 East 47 Street
New York, NY 10017

Personally, if you're not going to do a lot of research and head scratching (as I must do), I wouldn't buy copies of all this technical stuff. Some of it really gets deep! Unfortunately some of it is also obsolete in today's technical world. (Did you know that FCC Part 73 has *two* standards for TV–one for color and another for black-and-white). If you just want to look at the standards, spend a few hours in the library of a university with an engineering school near you.

BY THE WAY–

Did you know that RS-170A, as a standard, has yet to be adopted. According to the ***1990 Catalog of EIA & JEDEC Standards & Engineering Publications,*** it has been proposed since November 1977. Another reference I found said that it had been originally proposed in 1975. State of the ark?

Q: What are "NTSC Standards?"

A: The NTSC was a committee, no longer standing, that existed in the fifties to develop the color encoding standards for the USA. NTSC took the then-existing black-and-white "broadcast" standards and added color.

Back when I was but a wee babe, there was a big hubbub about "compatible color." This meant that the "new" color signals could be watched on a black-and-white TV set and that the "old" black-and-white signals could be watched on a "new" color TV set. That compatibility was the doings of the NTSC acting in response to discussions about color standards with RCA, CBS, Color Television Incorporated, Paramount Television Productions, Inc., and Chromatic Television Laboratories, Inc. RCA won.

(If you're a history buff, you can read about the intrigue behind the NTSC standards in ***Federal Communications Commission Reports*** Volume 41 (available in law libraries and from the Government Printing Office.) It's almost like figuring out whodunit in an Agatha Christie mystery!

The easiest way to get to the cold technological heart of the NTSC specifications is to look in Volume III, Part 73 ("Radio Broadcast Services"), Subpart E ("Television Broadcast Stations"), Paragraphs 682 ("Transmission standards") and 699 ("Engineering charts") of the ***Federal Communications Commission Rules and Regulations***. (Getting there is half the fun!) You can find this stuff in large libraries and in libraries that are federal depositories of information. You computer-types out there—you know who you are—get a copy of those specs and read them. It's at least five parsecs between "analog television video" and "analog computer video!" If you're going to try to sell to the television market, get your act together!

(Did you know that your VHS or ¾" VTR outputs are "NTSC-**Type**" signals? "RS-170" is "NTSC-Type," "RS-170A" is "NTSC." Watch for the "type" four-letter word on your specification sheets if you're worried about such things!)

Part III

Systems

General System Issues

Q: How should I route the cables in my equipment racks?

A: Keep the cables carrying incompatible signals separated by at least six inches. Usually I like to put shortened power cables on only one side of the rack, plugging directly into a vertical plug strip. On the other side of the rack are all the signal cables, grouped into compatible-signal bundles. There are basically four types of incompatible signals:

- Video, control, and R.F.
- Low-level audio (less than -4 dBm, like microphones)
- High-level audio (greater than -4 dBm, like "line level")
- Power

If the cables MUST cross (and they often do) they should cross at a 90 degree angle to minimize interference and cross-talk.

Q: What is the most important aspect of television?

A: Communication. What you say or show is a small part. The quality of your pictures and equipment is a small part. The bottom line of your production budget is a small part. *YOU MUST EVOKE THE EMOTION AND COMMUNICATE THE IDEA.* Don't cloud the big picture in favor of any little bit of the picture.

Q: What is brightness?

A: Brightness is the overall illumination level of a television display. It helps to think of brightness as the blackness of black picture details.

Pedestal and blanking adjustments to the video signal also affect brightness.

Q: What is contrast?

A: Contrast is the ratio of brightness of white picture areas to black picture areas. With brightness adjusted for blackness of black, contrast adjusts for the whiteness of white.

Gain and luminance adjustments to the video signal also affect contrast.

Q: What is color saturation?

A: Color saturation describes the amount of white that is present in a picture area. Saturation adjusts whether a hue will be fire engine red or pastel pink.

Chroma amplitude adjustments of the video signal affect color saturation in the picture.

Q: What is hue?

A: Hue describes the perception of tint of a picture area. Hue adjusts whether green is green or whether green is yellow on the display.

Chroma hue, burst phase, and chroma phase adjustments of the video signal affect color hue in the picture.

***Q:** When and for what sources do I need to add a frame synchronizer or a time base corrector? When can I simply provide a sync or genlock signal to the device? –Greg Rodgers, Atlanta, GA*

A: [Included in Greg's letter was a drawing (Figure 16) of a presentation system for consideration in developing the answer. Examining the drawing, we have several sources into one routing switcher. The routing switcher is capable of switching during the vertical interval. Monitoring, video projection and a production video switcher are connected to two routing switcher outputs.]

In order for a vertical interval switcher to operate properly, the vertical interval of each of the sources must arrive at the switcher inputs at exactly the same time. The signals will arrive synchronously at the switcher input if the following

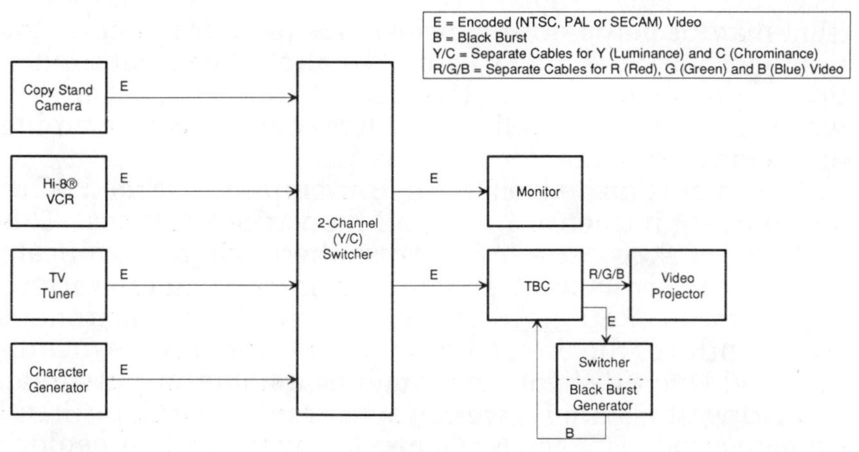

Figure 16
Original System Design

two conditions are met:

1. The various sources must be driven by a common sync reference. Only one metronome (sync reference generator–frequently a "black burst" generator) is allowed to provide the timing reference in any given system.
2. The delays encountered by the timing information must be equal among the various sources. The clicks from the metronome must arrive at the input to the switcher at exactly the same time.

If the source of the video signal is creating a new video signal, the operation of the signal creator can be easily and economically locked to the central sync reference. Video signal creators include cameras and character generators. Usually, these creators have genlock inputs to lock operations to the central sync reference.

If the source of the video signal is recording, processing or receiving a video signal from another piece of equipment (that may or may not be synchronous to your system), the timing of the video signal cannot be altered without sophisticated processing. Video tape recorders, TV tuners, microwave receivers and satellite receivers cannot alter the timing of the video signal.

The system detailed in Figure 16 attempts to use the TBC as a stabilizing influence among all the various sources. This will not work as expected because there will be significant sync timing errors induced by switching among the various inputs at the Routing Switcher. Once the asynchronous switch induces the timing error, the TBC (or even a synchronizer) will have difficulty maintaining a stable output signal.

Looking at Figure 17, we can synchronize the Copy Stand Camera and the Character Generator by using their genlock inputs. (They create new video signals.) The Hi-8® VCR requires the time-altering capabilities of the Time Base Corrector. (The VCR does not create a new video signal. Its playback signal was originally created by a camera or character generator. Notice that the new timing reference is provided the TBC to serve as the new metronome.) The TV Tuner requires the time-altering capabilities of a Synchronizer. (The TV Tuner does not create a new video signal. Its output signal was originally created by a camera or character generator at the TV station or network. Notice that the new timing reference is provided the Synchronizer to serve as the new metronome.)

The difference between a Time Base Corrector and a Synchronizer is slight. Basically, a TBC provides horizontal scan -by- horizontal scan correction–the kind of correction frequently required by tape playback signals. A Synchronizer provides vertical scan -by- vertical scan correction–the kind of correction required by non-synchronous signals that are otherwise stable.

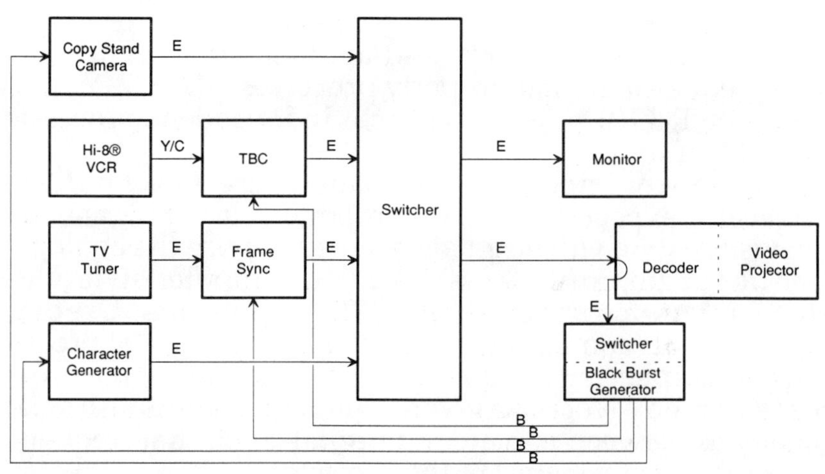

Figure 17
System Design Changed for Synchronous Operation

The sources in Figure 17 are now capable of synchronous operation and can now be timed so that the vertical intervals (actually the vertical sync pulses) arrive at the input to the routing switcher at exactly the same time.

Now let's take a look at some ways to maximize picture quality.

There are two basic rules to maximizing the picture quality of a presentation system:

1. Maximize the number of video signal "components" used. (I'm using the term "components" in quotes here because I'm liberalizing the use of the term to include "Y/C," "Y/ R-Y/ B-Y," "Y/I/Q," and "R/G/B." Y/C is not usually considered a true "component" video signal configuration, but it's a darn sight better than encoded composite NTSC video signals.)
2. Maximize the number of pieces that use video signal "components." A piece of equipment that properly processes Y/C will have two independent processing channels (one for Y and another for C). A piece of equipment that properly processes Y/R-Y/B-Y, Y/I/Q or R/G/B must have three independent processing channels.

One piece of important information here–most multiscan projectors expect an R/G/B component input signal. For television-style video signals, we usually use the analog R/ G/B input ports (if available). For computer-style video signals, we often use the digital (TTL) input ports. Use of the digital inputs sometimes means that special digital processing and amplifying equipment is necessary to get the projector to properly work in a given situation. There is also a need to switch between analog and digital–a job that frequently requires remote control of the projector.

Another piece of important information here–some projectors cannot tolerate any change in sync timing. When a sync discontinuity occurs to a light valve projector (all of which I am aware are manufactured by General Electric), a ragged blob of oil floats across the picture. With this type of projector, it is best to have a dedicated frame synchronizer so that there can be no sync timing discontinuities input to the projector.

What does this all mean when it comes to designing a presentation system? Figure 18 shows a "better" design approach to a presentation system. Figure 19 shows a "best" design approach to a presentation system. (Sounds kinda like a Sears Roebuck catalogue, doesn't it?)

To design the system for maximum picture quality, let us first look at the heart of the system–the input switcher. All television-style video signals must flow through this switcher. This basically means that this switcher must be capable of three-channel true "component" operation to maximize the quality of the projected picture. (We will use remote control of the circuitry inside the projector to switch between television-style video and computer-style video.)

Once we have three-channel "component" capability in the

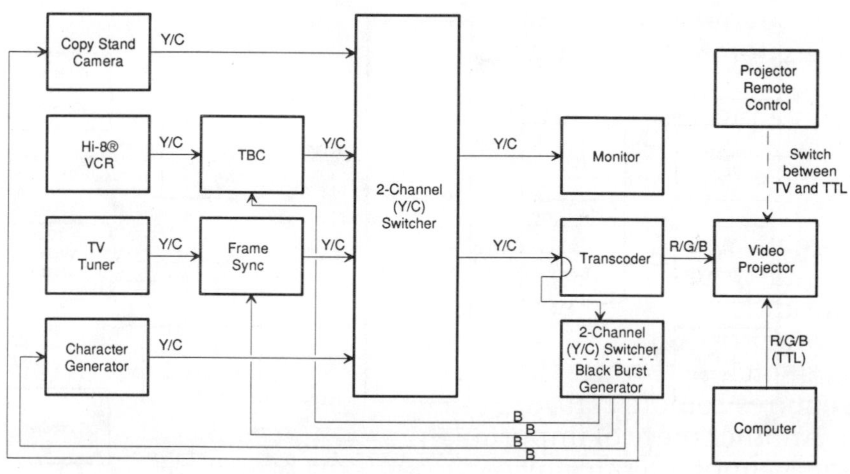

Figure 18
System Design Changed for Better Video Quality

projector and the switchers, then we need to worry about the signals that are input to the switcher. Nominally, all of them should be three-channel component sources.

Now we must choose *which* three "component" signals to use. The *de facto* television standard for "component" is Y/R-Y/B-Y. Most television equipment is designed to handle those "components." If the standard component complement is not Y/R-Y/B-Y, it's an easy and inexpensive process to convert.

R/G/B signals are more technically difficult to handle than Y/R-Y/B-Y signals. The signals in the R/G/B format require a wider bandwidth and can withstand a smaller amount of noise. If attempting to pass R/G/B through the switcher, get the highest quality switcher that you can afford. If the standard component complement is something other than R/G/B, conversion will be necessary.

Y/C signals should be converted into the selected components. Thank goodness, this is a relatively simple and low-distortion process.

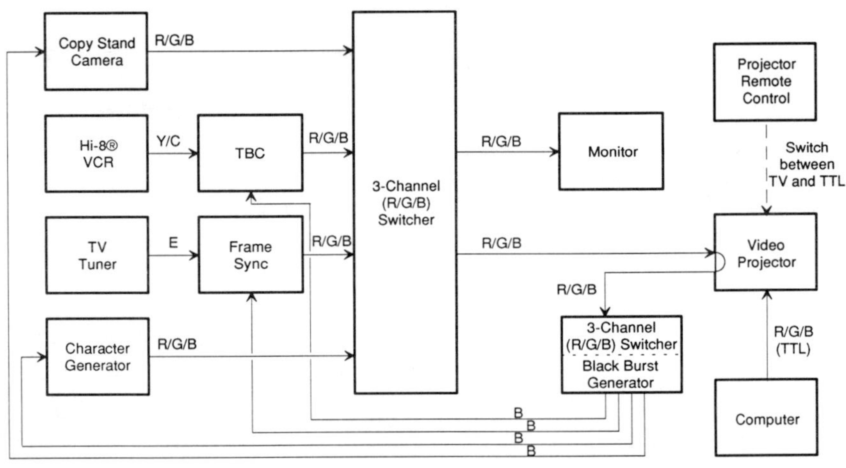

Figure 19
System Design Changed for Best Video Quality

NTSC signals should be decoded into the selected components. This is necessary to make sure that we can seamlessly switch among the various signal sources.

Once we have the right "components" into the right switcher, it's a relatively easy process to switch among the various television-style video signals; however, it's a different matter when it comes to switching computer-style R/G/B digital "TTL" signals. TTL signals have a very fast "rise time." If this rise time is lengthened, the resulting picture will suffer, sometimes to the point of producing no usable picture at all. Long cables (greater than about ten feet) induce distortions in TTL signals, requiring special distribution amplifiers and line driving amplifiers to maintain signal quality. For the same reason, the same system that switches and processes the television-style video signals cannot be used to switch and process the computer-style video signals.

So here's what we've got:

1. Subsystems to convert from the various television-style video signals to the chosen standard "component" video signal used in the presentation system.
2. A subsystem to switch and process television-style "component" video signals.
3. A subsystem to switch and process computer-style video signals.
4. A subsystem to project both television-style and computer-style video signals.

Despite all this high-tech technology, there will still be those instances where an overhead projector or slide projector are required by a particular presentation. Remember that low-tech is always going to be an alternative to presenters with technophobia.

It ain't cheap (or easy) to do it right!!!

Fractured Definitions

vid·eo \'vid-ē-ō\ *n.* any system that displays a picture regardless of quality. electronic signal containing picture information regardless of scan rate. recording frequently accompanied with incomprehensible audio.

film·ing \'film-ing\ *v.* the making of a video.

ed·it \ed-ət\ *n.* a program transition point used to cover-up unusable production sequences. decision to let a horse say the lines. *v.* assembling a video to match incomprehensible audio.

***Q:** In shooting overseas I have experienced a flicker in the video when using NTSC equipment. The flicker occurs in interior shoots (either lit or using available light) and in outside exterior shoots using available light at night. This flicker does not appear in similar situations when the video was recorded using PAL or SECAM equipment and then standards converted. I assume this has to do with the difference in power cycle between European and American systems. Is there some way to eliminate this flicker (through adjustment of the camera's electronic shutter, etc.)? — Edward J. Jurewicz, Indiana University, Bloomington, Indiana*

A: You're on the right track. The flicker that you see is the difference between the 59.94 fields-per-second American television scanning rate and the 50 pulses-per-second of light intensity. (Many lighting fixtures go "-off-on-off-on-" at twice the power line frequency. In North America, the lights flicker 60 × 2 = 120 times-per-second. In Europe, the lights flicker 50 × 2 = 100 times-per-second.) The difference between television scanning and light flicker shows up in the picture as a "beat" between the two frequencies.

The way to get rid of the flicker in the picture is to match the television standard with the power line frequency. You can change to European standard 50 field-per-second equipment while in Europe (when in Rome . . .) *or* you can change the power frequency applied to the lights (with an inverter or generator) to match the American standard. The conversion from European-to-American television standards has been specifically designed to reduce the amount of flicker in the finished picture.

I know of no way in which a camera can be modified or adjusted to eliminate the flicker. Adjustment of the shutter simply adjusts the duration during each field that the camera is sensitive to light–the field rate stays the same. Perhaps, with a tube camera, the target controls could be adjusted to make the tubes laggy, but that would smear any moving objects in the scene and would require some technical support to misadjust and readjust the camera.

Q: What is power?

A: pow·er \'paủ-r\ *n.* **1 a:** anything that is short supply when attempting to light for field production **b:** what the house electrician has in his little finger when the lights go out **2:** phrase seen in a cartoon bubble during a fight in a Batman episode **3:** the inverse of the rate at which work is not accomplished **4:** electricity converted to light when a 15 Watt Production Assistant gets plugged into a 100 Watt circuit.

(Hint: Definition 3 really *is* correct, if you use Reverse Polish Notation!)

Q: Why shouldn't I use an extension cord to permanently power my equipment?

A: The National Electrical Code (*NEC*) says that extension cords are only for "temporary" use. Any equipment that needs power for a period greater than 90 days *must* be powered through wiring permanently installed by an electrician. (The 90-day period is for "... Christmas, decorative lighting, carnivals, and similar purposes." Of course, every television facility is a carnival lasting more than 90 days!)

The concern here is twofold. First, the conductors and insulation used in the extension cord may be unable to support the load you want to impose. If there is too much electricity demand made through undersized conductors, they will heat up more than normal (possibly to the point of causing a fire). If the insulation is of a material that will not withstand high heat, it will melt. (As an experiment, you may wish to feel the heat coming from an extension cord after a high load has been on for thirty minutes or more.)

The circuit breakers in your electrical closet were sized during the design of the building to protect the permanently installed wiring going through the ceilings and walls to the duplex receptacles. The size of the wire, the material used in the conductors, and the insulation all conspire to establish the "rating" of a particular wire. (Bet ya didn't know that these circuit breakers are there to protect the wires, did ya? ... Huh? ... Huh? ... Huh?)

There is no guarantee that the conductors in an extension cord are sized to carry the same amount of electricity as the wiring in the wall. You can buy a 16AWG (American Wire Gauge), or a 14AWG, or ... (the lower the number, the bigger the conductors and the higher the load capacity). You can also buy a wide variety of types of insulation from neoprene, through vinyl, to rubber. (You thought these materials were only for birth control. Tsch. Tsch.)

Too much electricity for the chosen extension cord yields too much heat and a possible fire!!! The extension cord becomes the weak link–the current-limiting circuit breakers are sized for the "stronger" permanent wiring.

The second concern here is that the voltage delivered at the output of the extension cord will be too low. The voltage at the wall duplex receptacle is specified in the *NEC* to be within 95% of the voltage arriving at the system. What this means is that up to *five percent* of your electric bill might be being used to heat the conduits in your building.

If the extension cord is too long for the size of conductors, you can easily shed too much voltage for reliable equipment operation. This problem is a particularly sticky wicket when you plug an extension cord into an extension cord into . . . By the time you get to the end, the voltage has dropped. This forces the equipment to consume more current. More current imposes a greater load on the wire. The wire heats up.

To find out more information about all this stuff, there's help in your local bookstore and library. The ***NEC*** is the horse's mouth, but the ***NEC*** itself is a virtually unreadable bureaucratic nightmare. Fortunately, there are easier-to-read interpretive books that are well worth a comprehensive thumbing. McGraw-Hill publishes the ***National Electrical Code Handbook*** that gives a detailed prose description of the ***NEC***. Both books are widely available in bookstores. (Be sure to get the *latest* edition of the ***NEC*** or ***Handbook,*** for the specifications change dramatically every few years.)

A word of warning here–not all governmental entities have adopted the ***NEC*** as their minimum standard guideline. It's best to check with your local government to find out what building code and electrical code standards *they* use–there's a bunch of other standards out there.

Why is it important to follow the guidelines of your municipality? Insurance. If you use a facility in a manner that violates building or electrical codes, *you* are liable for damages in case of accident or fire. Some insurance companies won't pay off if you cause damage by willfully violating the minimum acceptable standards of your community.

Q: Why should I not tightly bundle my power cords?

A: Under some conditions, tightly bundling power cords into a small knot can cause a fire. It *is* unusual, but it has happened. Heating of the coiled-up power cord occurs because the coil of wire acts as an inductor that consumes electricity (even when the power switch of the equipment is turned off). The amount of heat created is dependent on the size of the coil, the sizes of the conductors in the coil, the number of turns in the coil, the insulation around the conductors, and the proximity of the coil to any "ground" or "neutral" potential (like an equipment chassis). The other danger is that as the insulation ages under heat, it will be either melt or become brittle with age.

Usually, the problem has occurred with two wire power cords on kitchen appliances where a homemaker has tightly bundled the cord so that there is minimum power cord exposed between the appliance and a nearby duplex outlet. I don't *think* there could be a problem with 3-wire cords just because of their size and bend radius, but who knows? If in doubt, crawl around to the back of your equipment and feel around your coiled-up power cords for any heat being created. You've got too much money concentrated in one small area to take any chances.

System Resolution

Q: Why should I buy a camera with 800 TVL of horizontal resolution when my VTR can only handle 260 TVL?

A: Remember from past articles that the higher the number expressed as horizontal resolution, the sharper the apparent image. If the signal starts out sharp, it can withstand considerable distortion before it becomes objectionable in the picture. If the signal starts out fuzzy and dull, it can withstand considerably less distortion.

In general, it may be helpful to think in terms of sharp edges on objects in the picture. A piece of equipment with lower resolution will simply reduce the sharpness of those edges. If you start out with sharp transitions, the transitions will remain sharper than if you had started out with dull, indistinct edges.

More about resolution:

With all the doublespeak about picture quality, there's one factor that always arises in discussion–resolution. Resolution specifications are usually given more weight than contrast ratio, signal-to-noise ratio and geometric accuracy in expressing overall picture quality, although they contribute heavily to the perceived quality of a picture. There are so many questions about what resolution means, what factors affect resolution and how is resolution measured that I figured that it's about time to help you resolve resolution.

First, our definition; with all due respect to Webster,Funk, Wagnels, et al., in this article I'm going to define resolution as "the ability to clearly discern details in an image." The smaller the details that can be seen in an image, the sharper the image and the better the perceived resolution of the image.

There are any number of ways in which picture resolution can be affected. Limitations can be placed on resolution by virtually any piece of equipment or cable in the entire system, including improper scene lighting, smoke-filled rooms, a hangover or other outside influences.

With all these ways in which resolution is affected by the various pieces of equipment, there are an equal number of ways in which resolution is measured and described. A portion of the terms are derived from the old tube-type television technologies. The remainder of the terms are derived from the newer solid-state computer technologies. It is not always possible to directly convert from one method of describing resolution to another.

TELEVISION RESOLUTION

Resolution of the television system is usually found measured in one of four ways–TV Lines, rise time, bandwidth or dot pitch. Picture details along the horizontal axis are more difficult to achieve than along the vertical axis of the picture. Most specifications refer to "limiting horizontal resolution" and assume that vertical resolution can't get any better than the number of active horizontal scans can allow.

Limiting Horizontal Resolution measured in TV Lines (TVL)

Limiting horizontal resolution is the number of individual black areas that can be seen on ¾ of each horizontal scan. The white areas between the black areas are equal in size to the black areas. The greater the number of individual lines (called "TV Lines"), the sharper the detail in the picture. (Note that horizontal resolution has no direct relationship to the number of horizontal scans–525 in North America. That's the limiting *vertical* resolution.)

Limiting horizontal resolution is measured with a "Resolution Test Chart" (a portion of which is shown in Figure 20) and a high-resolution picture monitor (usually a black-and-white picture monitor). The vertically converging lines at the top-center and bottom-center of the chart demonstrate the limiting horizontal resolution of the system. There is a point past which individual black lines cannot be seen in the picture. This is the limiting horizontal resolution of the system being measured.

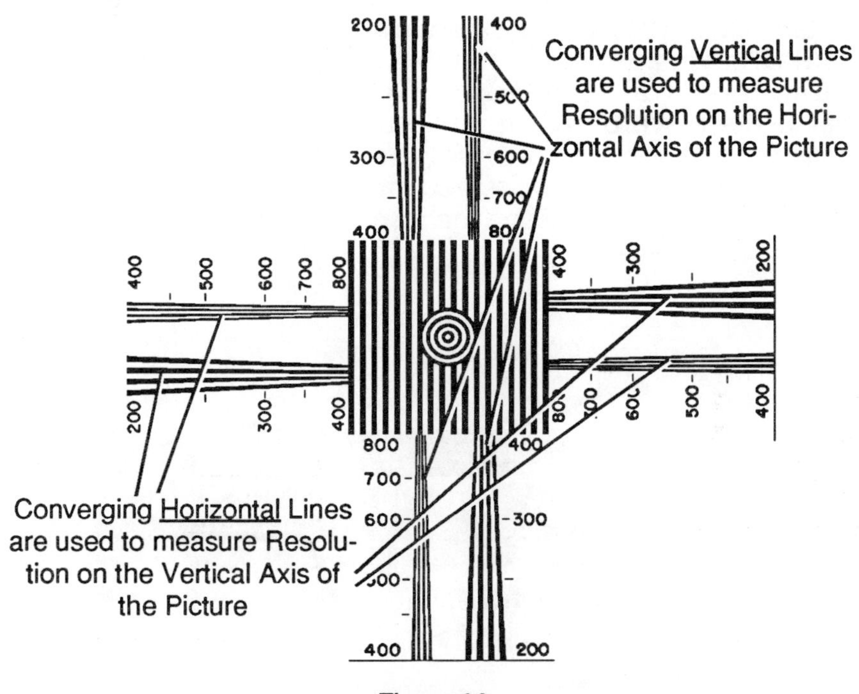

Figure 20

Measuring Resolution in TV Lines with a Resolution Test Chart (Only the center of the chart is shown)

The number printed next to the point where the lines converge represents the limiting horizontal resolution expressed as the number of individual black lines that can be seen in ¾ of a horizontal scan. The chart must be calibrated to the television frame for the numbers to be accurate. Calibration is achieved when the points of the white wedges at the chart edges fall on the edges of the scanned area of an underscan monitor.

Limiting Horizontal Resolution Measured as Rise Time

Horizontal resolution is sometimes measured in the length of time it takes for the video signal to transition from a "low" voltage from a dark image area to an adjacent "high" voltage from a light image area. As shown in Figure 21, the time spent in transition is called the "rise time." The amount of smearing that takes place between adjacent picture details is being measured; the shorter the rise time, the smaller the smearing and the sharper the perceived image. Figure 22 shows a perfectly sharp transition (the impossible dream).

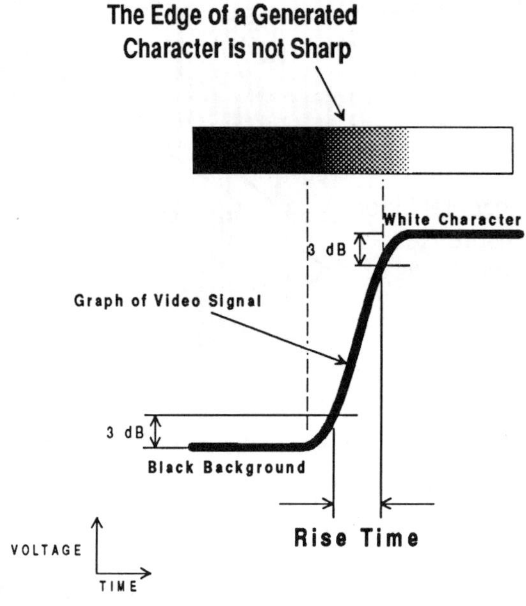

Figure 21
Measuring Rise Time

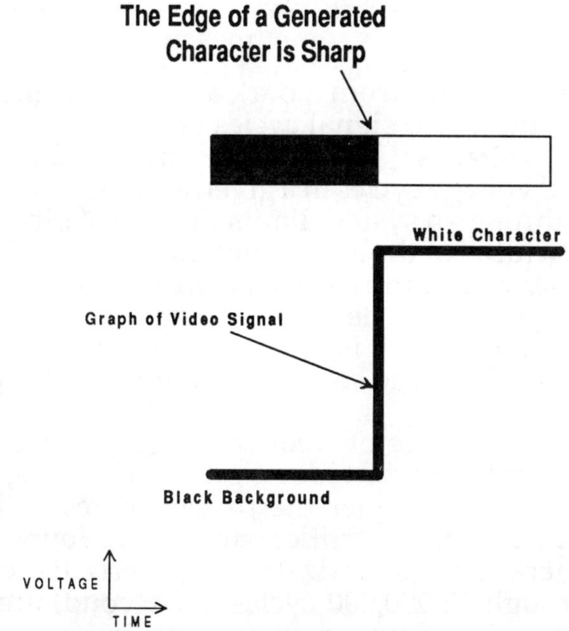

Figure 22
Zero Rise Time of a Perfectly Sharp Character
(Impossible to Achieve)

This "rise time" is usually measured in nanoseconds—0.000000001 seconds. Rise time can be electronically measured with an oscilloscope. Normally this specification of resolution is found associated with character generators where the edge sharpness of the electronically generated characters is important.

It is possible to convert from resolution specified in terms of rise time into TV Lines. Divide 26050 by the rise time (in nanoseconds) to come up with the equivalent number of TVL resolution. (It's a close approximation.)

Limiting Horizontal Resolution Measured as Signal Bandwidth

As the picture goes from a dark area to a bright area (and vice versa), the video signal cycles between "low" voltages and "high" voltages. The practical limit to the number of video signal voltage cycles in a given amount of time that can be passed through a system limits the size of picture details. The "bandwidth" of the electronics through which the video signal passes can limit the perceived resolution of the displayed images. (When describing imaging devices, the term "depth-of-modulation" is frequently used to describe the bandwidth of the video signal emanating from the imaging devices.)

The wider the range of frequencies (expressed in cycles-per-second, or Hertz) of video signal that can pass through the electronics, the higher the perceived resolution of the image. Typical video amplifier bandwidths found in professional cameras are 50 Hertz-15 MegaHertz (50 cycles-per-second through 15,000,000 cycles-per-second) area. Lesser cameras may have upper frequency limits as low as 5 MegaHertz.

Figure 23 shows a typical multiburst pattern that may be used with an oscilloscope to measure video bandwidth. Figure 24 shows how a multiburst signal is distorted as it goes through a typical heterodyne recorder (like ¾" and VHS). Notice that the 3.0 and 4.2 MHz bursts are missing entirely. This means that sharp edges in the original picture will be dulled as the signal passes through the recorder.

Notice that the 1.0 and 2.0 MHz bursts in Figure 24 have been attenuated. This means that the whites don't get as white and the blacks don't get as black as they did in the 0.5 MHz burst. This reduction in contrast will be perceived as a reduction in resolution. It turns out that contrast ratio and the ability to resolve picture details are inextricably intertwined. After all, with a contrast ratio of zero (with the lights out), you can't see *any* picture details. This is one reason why a properly lit scene appears sharper than a flatly lit scene.

Increasing Frequencies

Increasing Resolution

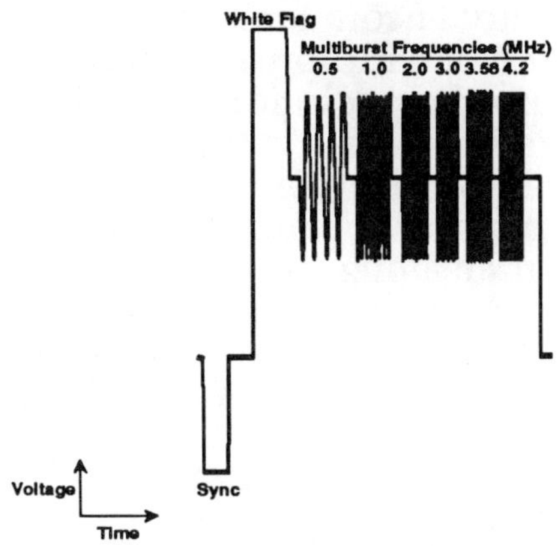

Figure 23
Typical Multiburst Signal
(Shown with North American Frequencies)

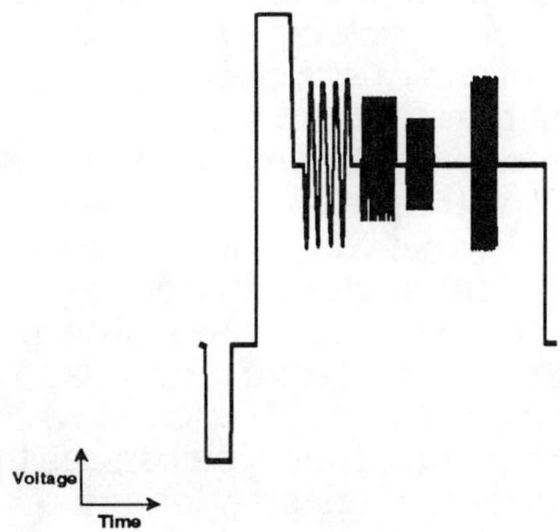

Figure 24
Typical Multiburst Signal after Passing
Through a ¾" or VHS Recorder

Depth-of-modulation is measured with an oscilloscope and a depth-of-modulation test chart. Expression of the depth-of-modulation is usually encountered in imaging technology and not encountered in widespread practice.

As a rule-of-thumb, the resolution along the vertical axis of a television picture is limited by the number of horizontal scan paths appearing in each frame of the picture. (In North America, it's the magic 525 horizontal scan paths in each frame.) The resolution along the horizontal axis of a television picture is limited by the number of individual picture details that can be seen on *each* horizontal scan path.

It is possible to convert between bandwidth and horizontal resolution measured in TVL. With the North American System-M scanning standard, you get 80 TVL resolution for each 1 MHz of bandwidth.

Color Picture Tube Dot Pitch

Inside the transparent face of each color picture tube, three types of glowing materials (affectionately called "phosphors") are coated. One type of phosphor glows red when struck by electrons. Another type of phosphor is made up of chemicals that glow green when struck by electrons. A third type of phosphor glows blue when struck by electrons. Three electron beams are generated in the neck at the rear of the picture tube and are carefully controlled to strike the phosphor-coated areas that glow the proper colors. (A black-and-white picture tube has only one, continuous coating of phosphor and is not subject to dot pitch limitations.)

In this system, the red, green and blue phosphors are excited to glow in various relative amounts to create all the colors in the rainbow (including black and white). If you get up very close to the picture tube, you can see the individual areas where the various colors are glowing.

It takes a complete set of three phosphor areas; one red, one green and one blue; to display all the colors that can potentially appear in a given picture detail. The minimum size of picture tube area that can accurately describe any picture detail must contain a complete set of three of the individual phosphor areas. As shown in Figure 25, the minimum horizontal distance that contains three colors of phosphor is called the "dot pitch" of a given color picture tube. Dot pitch is usually expressed in millimeters (mm). The smaller the distance, the greater the number of individual picture details that can be seen in the picture.

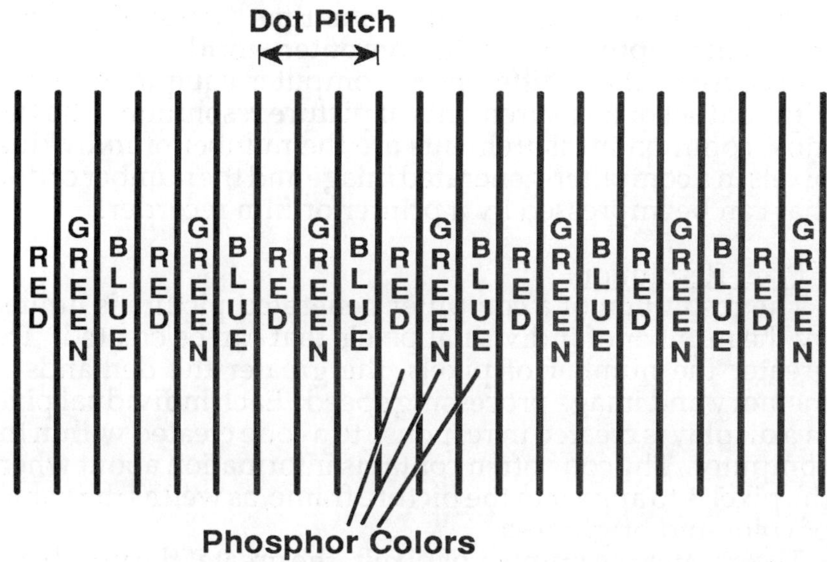

Figure 25
Measurement of Dot Pitch

COMPUTER RESOLUTION

Computer systems, left to their own design, would create pictures with a different technique than television systems. They would ideally operate in a manner similar to the human eye. Instead of scanning the image, our eyes dissect the image into individual bits and pieces. The rods and the cones of the retina are individual light sensing elements, each of which is connected by an individual "wire" in the optic nerve to the electrochemical computer we call a brain.

Instead of scanning a real image and transmitting it down a few cables (as is done in television), computers could ideally create the color in individual picture elements (pixels) and connect each creation element to an individual display element via an individual cable. It's the difference between the slower one-piece-after-another "serial processing" of information and the fast all-at-one-time "parallel processing" that is optimum for the computer world.

Because of these differences, computer equipment uses a different set of measurements of picture resolution. The two most common measurements are the number of individual pixels in a computer-generated image and the number of dots that can be impressed by a printer or film recorder.

Pixel Resolution

The resolution of a computer generated picture is limited by the number of individual pixels that can be created. The greater the number of pixels, the greater the demands of memory and image processing speed. Each individual pixel in a display is created in response to a code created within the computer. The code often contains information about where the pixel is to appear in the picture frame, as well as the values of color and brightness.

The greater the number of pixels, the greater the number of memory locations that must be provided by the computer. The greater the number of colors and gray levels, the greater the number of memory locations that must be provided by the computer. With a limited amount of expensive memory, many systems offer the option of reducing the number of pixels in the picture in trade for more colors to appear in the picture. In the other direction, fewer colors in the picture often frees enough memory to allow an increase in the number of pixels.

As shown in Figure 26, the number of pixels in a given picture is expressed as the number of pixels on the horizontal axis of the picture -by- the number of pixels on the vertical axis of the picture. For example, with a 1024 X 768 picture you could squint and count 1024 individual pixels horizontally and 768 individual pixels vertically in the picture. The total number of pixels in the entire picture is 1024 × 768 = 786,432 individual pixels in a closely-spaced array.

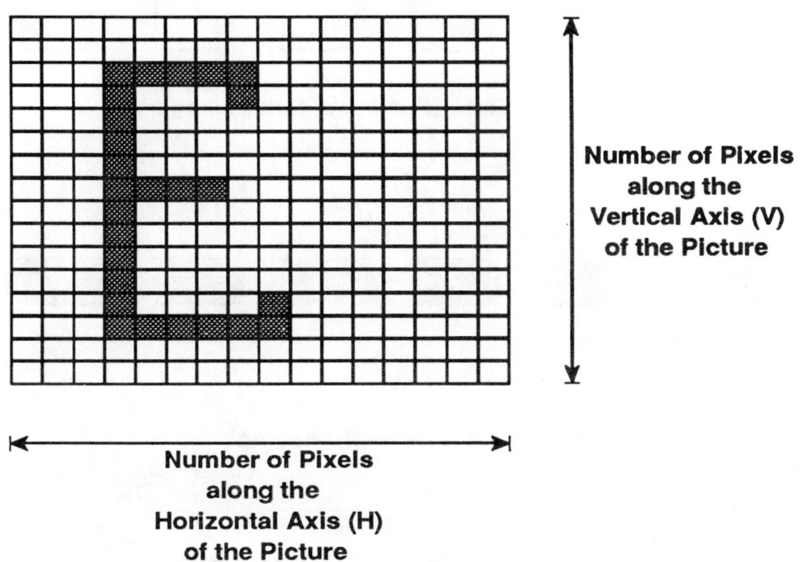

Figure 26
Measuring Resolution in Pixels

But wait, couldn't you assume that half the 1024 pixels are black and the other half are white to come up with 512 black lines on an entire horizontal scan and 384 black lines on ¾ of a given horizontal scan. Doesn't that give us the equivalent of 384 TVL limiting horizontal resolution expressed in television terms? Not really. As shown in Figure 27, the shape of the sharply-edged pixels versus the Gaussian-edged TV Lines are different and produce different measurement results. Remember we're talking new solid-state technology versus old tube technology. There's no easy way to accurately convert from pixels to TV Lines.

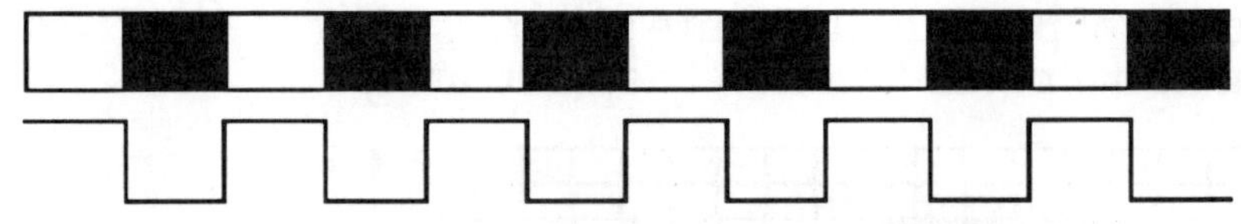

a) Solid-State (Pixel -by- Pixel) Imaging

b) Tube (Gaussian) Imaging

Figure 27
Comparison between Solid-State
and Tube Resolution Measurement

Solid-state "CCD" imaging chips used in newer television cameras use chips that image with discrete pixel-sized imaging elements. With these cameras, it is more appropriate to talk about pixel resolution instead of limiting horizontal resolution. The smearing between picture elements was much more of a factor of tube imaging than solid-state imaging.

Dots-per-Inch Resolution

Many computer output devices measure resolution in terms of the number of individual dots that can be discerned in a given length. Figure 28 shows how an array of black dots (black ink dots on white paper) creates an image. For laser printers, the typical limiting resolution is around 300 dots-per-inch (dpi) along the vertical and horizontal axes of the paper. Professional "typeset-quality" "image setters" have limiting resolutions around 1024 dpi and up. The greater the number of dots, the more storage area and memory is required. The greater the number of dots, the less ragged will be the appearance of any diagonal elements on the page (particularly on some letters like "A," "X," and "V").

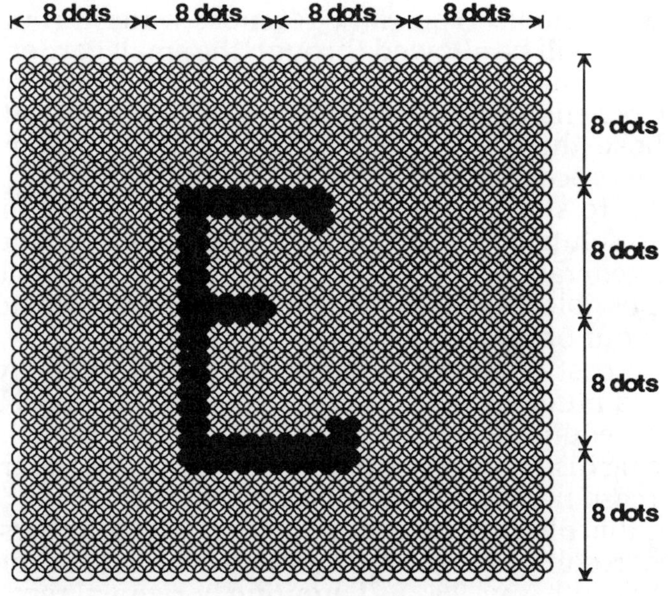

Figure 28
Dot Resolution

The other factor of printer resolution is the ability to accurately reproduce "half-tone" photographic work. Generally, the higher the resolution, the more natural the reprinted photograph appears. Remember that printers typically have only black ink to splatter on paper, so the relative amounts of white-paper -to- ink-splattered paper establishes the ability to reproduce shades of gray.

SYSTEM RESOLUTION

Now let's look at some of the more confusing aspects about resolution. First, it is true that the resolution through a system of individual pieces of equipment is limited by the weakest link (lowest resolution) component in the system.

It's somewhat akin to trying using a garden hose to replace a large water main. You can't get more water through the system than will be allowed through the small garden hose, regardless of the size of the remainder of the pipes. There is a hidden benefit to using the larger pipes before and after the garden hose–there will be less turbulence in the water at the end of the process.

It is best to start out with a picture that has the highest feasible resolution. This means a high-quality camera or graphics generator. This will give you a chance to provide the highest possible quality of system picture output instead of limiting your options at the start.

It is also best to maintain the highest resolution circuitry through as much of the system as you can afford. Higher resolution equipment will not limit the picture quality by introducing distortion-producing artifacts in the picture that will decrease perceived resolution.

Taking this philosophy a step further. The numbers used to express resolution are not absolute. An individual piece of equipment with, say, 260 TVL limiting horizontal resolution will not usually limit the entire system to only 260 TVL horizontal resolution. It will degrade the resolution, but not by that much! If you start out with a picture created by a camera capable of 800 TVL resolution and send it through a recorder capable of only 260 TVL resolution, the picture will still appear sharper than if you started with a camera capable of creating a 400 TVL picture.

Another common misconception concerning resolution specification is about what devices are included in the specification. With computer systems, it is often difficult to figure out whether the 1024 X 768 capability is associated with a computer display, a film recorder, a graphics card, a computer architecture, a television output interface, or what. For instance, some of the illustrations accompanying this article were scanned in monochrome at 300 dpi, edited on a 1024 X 768 by 256,000 color display with a 1016 dpi digitizing tablet, and printed on a 300 dpi laser printer. When interfacing a computer for use with a television system, it is often difficult to determine whether or not the promised resolution and number of colors are indeed output by the television interface. The only way around these problems is to read specifications carefully, ask questions and consider caveat emptor the credo of the television and computer equipment businesses.

Somehow, a simple concept like picture sharpness and detail has taken on a myriad of subtle connotations. Despite all the measurements, measurement units and specifications, it all boils down to what looks good to you. Don't let the numbers on a specification sheet deceive you–make sure you have a clear picture of the situation.

System Timing

Q: How do you "synchronize a system?"

A: The various pieces of television equipment that are interconnected to create a system must scan the television image at exactly the same time. If the various pieces of equipment do not scan at exactly the same time, the picture will roll or shift whenever the various signal sources (like cameras, character generators, VTRs, etc.). It is this scanning of the image that we must "synchronize" for proper system operation.

Electronic signals are distributed throughout the system to synchronize this timing operation. These "sync" signals serve as a metronome in the scanning process. There are several names associated with the sync signals used within a given piece of equipment. For example, operation of the various circuits in all tube-type color television cameras require the following "sync" signals:

(1) Vertical Drive–sent to "vertical deflection circuits" to time the scanning of the vertical axis of the picture;

(2) Horizontal Drive–sent to "horizontal deflection circuits" to time the scanning of the horizontal axis of the picture;

(3) Composite Blanking–sent to pick-up tubes to time the turning-off of the electron beams in the pick-up tubes during image scanning "retrace" from bottom-to-top and from right-to-left of the picture;

(4) Composite Sync—sent to the "sync adder" circuit and added to the camera output signal to tell other equipment what times the camera started horizontal and vertical scanning in the creation of its video signal;

(5) Subcarrier—sent to the "encoder" circuit to time ("phase") the operation of the color encoder circuit that combines separate red, green, and blue video signals created within the camera; and,

(6) Burst Flag (sometimes called Burst Gate)—sent to the encoder circuit to time the start and stop of the burst signal that tells other equipment how the camera created the color (chrominance) portion of its video signal.

If any of these six "sync" signals are missing, the camera will not create a reliable color video signal. If the camera were to lose one of these signals it could create the following problems:

(1) Vertical Drive—the picture collapses into a single horizontal line (because the camera is never told a time to start a scan on the vertical axis of the picture);

(2) Horizontal Drive—the picture collapses into a single vertical line (because the camera is never told a time to start a scan on the horizontal axis of the picture);

(3) Composite Blanking—a pattern of white dots will appear in the picture (where the active scans and the normally-blanked retrace scans intersect);

(4) Composite Sync—signal destination equipment will not "lock" to the output video signal (because it isn't told how the signal was put together in the first place);

(5) Subcarrier—color hue (if present) would be incorrect (because the color encoder circuits in the camera are never told a reference phase to use when creating the color portion of the signal); and,

(6) Burst Flag—the camera output signal will be black-and-white because a burst will not be present and a burst is required in an NTSC encoded color signal (because the camera is never told when to start and stop the creation of the burst).

A camera used in a system must have each of these signals synchronous with other video signal sources in the system. Each of these signals carries different information in the form of voltage pulses at regular intervals. (The subcarrier is a little different in that it forms a continuous "sine wave" instead of discrete timed pulses.)

Some notes about technology are appropriate here–the term "sync" is bantered about and given all manner of meanings and connotations.

"Sync" may refer to the concept of a central time reference (like a metronome) that commands the synchronous operation of a system.

"Sync" may refer to the signal that I (and most other engineers) call "Composite Sync."

"Sync" may erroneously refer to a signal used as the "reference" for genlock operation (like "black burst").

Sink may refer to "Titanic" operations.

Sink may refer to a receptacle in a kitchen.

OOPS! A little humor never hurts. (Very little.)

The delivery of the synchronizing information to a given camera (or other signal generating, recording, or processing equipment) is where there are a lot of unanswered questions. There are two basic approaches to synchronize equipment—what I call "sync lock" and genlock.

"Sync lock" was the original approach used to synchronize the various pieces of equipment that are interconnected for form a "system." A piece of equipment that uses the sync lock system will be connected to as many as six coaxial cables delivering sync signals (one for each of the six different types of sync signals). System timing (that we've talked about in earlier installments) required not only timing from one piece of equipment to another but also from one type of sync signal to all five remaining types of sync signals. All the sync signals are generated in one separate "master sync generator."

Genlock is the "new" approach (since the 1970's) of synchronizing each piece of equipment in a system that synchronizes equipment with one signal flowing through one cable. I think that genlock is the outgrowth of a system proposed by CBS Laboratories called a "Unipulse" system. (If anybody knows different, let me know.)

The key to understanding genlock systems is to realize that each piece of genlock equipment has a miniature "master sync generator" within it. The six sync signals output from this sync generator circuit are permanently connected to the appropriate circuits within the case of the equipment. On most types of equipment that use genlock, these signals are not available via connectors on the outside of the equipment, but they are present within the equipment. Thumb through the service manual for your camera (you *do* have service manuals for your equipment, don't you???) and you'll find all six types of these sync signals.

To make a piece of genlock equipment synchronous, it is the miniature internal sync GENerator in that equipment that needs to LOCK the timing of the generation of the six types of sync signals required for proper operation. If you lock the timing of the generation of the internal sync generator, you lock the timing of the equipment for synchronous operation.

One other big difference between "sync lock" and genlock

equipment are the signals used to synchronize the equipment. Systems using synclock techniques require the six sync signals to have the following peak-to-peak (bottom-to-top) voltages:

Vertical Sync	4 Volt	rectangular pulses
Horizontal Sync	4 Volt	rectangular pulses
Composite Sync	4 Volt	rectangular pulses
Composite Blanking	4 Volt	rectangular pulses
Subcarrier	2 Volt	continuous sine wave
Burst Flag	4 Volt	rectangular pulses

These are the voltages output from the various connectors on the back of separate sync generator equipment.

Now what about genlock. A genlock circuit is designed to accept any composite encoded color video signal as its timing reference. Most people prefer the term "reference signal" rather than "sync signal" when talking about genlock operation. The voltages expected by a genlock circuit are those associated with a composite video signal, namely:

Composite Sync	0.286 Volt (40 IRE rectangular pulses)
Burst	0.286 Volt (40 IRE intermittent sine wave)

The genlock circuit uses the timing of the sync pulses in the incoming reference signal as the times to generate pulses in the Vertical Drive, Horizontal Drive, Composite Blanking, Composite Sync and Burst Flag sync signals for use within the equipment. The genlock circuit uses the timing of the voltage peaks and valleys ("phase") of the burst in the incoming video signal when generating the Subcarrier "sine wave."

The sync and burst are all that are important to the genlock input circuit–picture content doesn't matter. The picture can be from a capped camera (which is all "black burst" really is), color bars, "Monday Night Football" on Saturday night (which I will never understand), or "Attack of the Killer Tomatoes," or taped replays of "Saturday Night Live," or . . . (Warp 5 out there, Mister Sulu!)

"Black burst" is a cheap signal to generate. It also has all the ingredients to properly serve as a genlock reference circuit. This leads to the Smithy corollary of economic reality:

Cheap + Everything Needed = Widespread Use

Notice that it takes considerably more voltage (up to 4 Volts peak-to-peak) to drive a sync lock system than a genlock system (0.286 Volts peak-to-peak). This means that a sync lock circuit will probably never see the small sync voltages available in a genlock reference signal.

Now for the confusion—some manufacturers mislabel the connectors on their equipment. Some connectors labeled "sync" are connected to genlock circuits. Some connectors labeled "reference" are connected to sync lock circuits. If in doubt, try whichever signal(s) is readily available. As long as it works, it doesn't matter what is "technically correct"—you won't "burn anything out." Like my extinguished ... errrr ... distinguished colleague Dick Reizner always sez—"It doesn't have to be, it just has to appear to be!"

Q: What's a tweaker?

A: tweak·er \'twēk-ər\ *n* **1:** most dangerous weapon in the arsenal of man **2:** screwdriver with a 3/32" blade size and a green handle frequently found in motion without any evidence of an intelligent controlling force and providing sufficient evidence to support a defense of justifiable homicide in a court of law **3:** controlling force behind same, with or without evidence of intelligence **4:** job security for qualified technicians

tweak·er *vb* one who tweaks

Q: Where in a video system is it important to have the system in H-Phase?

A: At the input to the production video switcher (sometimes mistakenly called the "SEG"—"Special Effects Generator"). If the sync pulses are not in time *at the input to the switcher* you will get a horizontal jump in the position of the picture as a dissolve or wipe is completed. (Some switchers allow you to switch among the various inputs with the preview bus, but the timing *must* be established at the input to the switcher.) This jump is the picture monitor's way of showing you that there has been a change in the timing of the horizontal sync pulses for a fraction of a second. If you don't see a jump, you don't have a problem. It's that simple.

If you were to try to measure the system timing at any point in the system other than at the input to the switcher, you will wind up with erroneous information. Because of different cable lengths and different amounts of delay in various equipment designs, the timing of the system will probably *not* be synchronous as you follow the synchronizing pulses as it progresses on its merry path.

The actual method of system timing depends on the equipment in the system. For the sake of argument, let's assume that each piece of equipment in your system is equipped with a "genlock" input. This means that each piece of equipment will have an "H-Phase" control. (If the equipment does not have a "genlock" input or if it does not have an H-Phase control, another method of timing must be used. That system of timing requires the introduction of delay in the sync line by lengthening the cable or by adding a "delay line.")

Now let's look at the switcher. If it has any special way of displaying H-Phase inequalities (like converging lines, etc.) in a special timing mode, skip the next paragraph–follow the switcher manufacturer's procedures, use that built-in capability and go on vacation.

If the switcher doesn't have a special timing display mode, does it have a "blanking processor" circuit? (You should look in the "Operator's Manual," the "Service Manual," or on the

"Specification Sheet" for this information.) *As we time the system, it is very important to find a path through the switcher that bypasses any blanking processing because it can easily work against us.* Normally if a blanking processor is in the switcher, it will be connected only on the output of the "Program Bus," bypassing the "Preview Bus." Some switchers have a switch that will allow the blanking processor to be turned off without affecting any other timing parameters.

Now, for the sake of argument, you are using the sync generator circuit within the switcher as your master sync reference. This means that you will probably be using a "black burst" signal that is output from the switcher, to which you will lock all your other equipment.

Now, let's put a color background generator circuit within the switcher. That video signal source must be timed just like any other video signal source (like cameras, character generators, etc.). The phase of the horizontal sync pulses from the color background generator must be timed at the input to each of the video signal busses in the switcher.

Let's make the video signal from the color background generator the H-Phase reference. (Usually we have no choice because there is no external adjustment for H-Phase of the switcher's color background generator circuit. The H-Phase and SC-Phase adjustment on the switcher does *not* directly make this adjustment–they are adjustments to time the switcher to yet *another* sync timing reference that may be input to the switcher's genlock circuit.)

Now, we simply time the remainder of the video signal sources, one-at-a-time, to be in H-Phase with the color background generator. You can do this by adjusting each video signal source until you no longer see the picture position jump at the end of a dissolve or wipe between the source and the color background generator. You can really see it if you have a cross-pulse display of the signal on a picture monitor that can be externally referenced to a constant sync source. You can also look for no horizontal jump in the graph presented on a waveform monitor (with the waveform monitor triggering on external sync), or you can use any built-in switcher H-Phase indication.

There are many, many variations on how to measure H-Phase inequalities and I have yet to find a method that consistently produces the most accurate results. Just remember what you're trying to do–time the horizontal sync pulses in each video signal source as they arrive at the input to the switcher.

Do not use color bars as you adjust H-Phase! Use any other picture except those that have no details. Color bars can really throw you off because the width of each of the color bars in each signal is adjustable from one piece of equipment to another. Unequal bar widths can make you think that H-Phase is off when it really isn't! (There really isn't anything wrong with using color bars, if you can pat your head and rub your stomach and count your beads all at the same time. It's just that it's a lot easier using other signals.)

On a slightly different subject, at the instant that you see a horizontal jump, you may also see a sudden shift of the hues of the colors in the picture. This is an indication that the subcarrier phase (S/C Phase) is out of adjustment.

Remember in all of your system timing adjustments and calculations, you are timing for phase coincidence at the input to the switcher. If you want to measure or detect the time (AKA "phase") difference, that's where you look.

***Q:** How can you introduce delay into a signal path?*

A: You can use cable. Lots of cable. Ever wonder why the control room raised floor felt spongy as you walked across it? One foot of Belden 8281 cable will introduce about 0.00154 microseconds of delay. One foot of normal RG-59 will introduce about 0.00130 microseconds of delay. Cable will introduce some distortion to the signal, particularly with long cable runs where equalization sometimes becomes a serious concern.

Or you can use one of the handy dandy delay boxes

available from any of several manufacturers. They're pricey (about $130), but they're a lot easier to handle than a boa constrictor made of cable. You can get 'em in any of several sizes, so you need to measure the amount of desired delay.

Q: What do I have to worry about when timing a component analog video system?

A: Here, we're talking about systems that support the "component analog video" (CAV) formats like BetaCam® and M-II. These systems use three cables to convey three different components a single video signal. One cable carries the luminance (Y) component of the signal. A second cable carries the red minus luminance (R-Y) component of the signal. A third cable carries the blue minus luminance (B-Y) component of the signal. Three separate cables are used, although they are sometimes enclosed in a single sheath.

In component operations, a delay *may* be induced in any of the three cables. Delays generally are induced by cable (the longer the cable, the longer the delay) and by processing equipment. If a delay is induced into only one of the three signals but not the remaining two (an unequal cable length), the components will not arrive at their destination at exactly the same time. This is a no-no. The outline of the lips may get there earlier than the lipstick on the lips, or worse.

Suffice it to say that in component analog video systems, we hold these truths to be self-evident: all delays are to be created equally. (Has a nice ring to it, doesn't it? As long as that darn bell doesn't crack again!)

Once the delays in all the signals have been equalized, you can time the component television system just like any composite television system. Remember that the object of this exercise to get all the Y, R-Y, and B-Y signals from all the video signal sources to arrive at the input to the switcher at exactly the same time. This is accomplished by inducing equal delay in the cables or by adjusting the H-Phase and S/C-Phase controls in the genlock circuit of the video signal source.

***Q:** How can I use the keyer or color background from my character generator and also use the generator as a downstream key source?*

A: The color background output from the character generator (CG) is a complete video signal with colored letters, backgrounds, and drop shadows. The key output from the CG is usually a white-on-black "cookie cutter" into which the color from a color generator circuit in a switcher can be placed.

These are two completely different signals with different intended usages that must be accomplished at different locations in a switcher. The video signal with color letters, etc. must be treated like any other video signal and must be timed with all other video signals input into the switcher. Once this signal goes through the various mix/effects busses, it gets delayed before it arrives at the downstream keyer.

If we try to dissolve between this delayed video signal and the downstream key cookie cutter, there will be a noticeable horizontal shift in the letters. The key output from the CG has not been delayed by the same amount as its color video signal.

To "fix" this problem, the key output from the CG must be delayed. The first step is to measure the amount of delay (as described in the previous question–sneaky, huh)? Measure the amount of delay between a character in the color CG signal and the key signal by using the video switcher to select one, then the other. Or you can simply use the downstream key over the color CG signal.

Now we know how much delay, how do we accomplish the delay of the key signal? This leads me to the next question

***Q:** How do I know that I'm maintaining SC/H phase correctly? —Ron Lambros, Rehoboth Baptist Church, Tucker, Georgia*

A: SC/H (subcarrier-to-horizontal) phasing is important in high-dollar editing systems and in broadcast facilities. There are several methods of measuring SC/H phasing, depending on the test equipment available. Some waveform monitors have special measurement routines and others simply have an LED indicator on the front panel to indicate proper SC/H phasing.

Where SC/H phasing is important is at the input to the video switcher. At that point, all signals must be in H-Phase, SC-Phase and SC/H-phase to eliminate position and hue shifts in the picture. Consult the switcher manufacturer's recommendations about the proper timing procedure for your model of switcher. Some switchers will use the preview bus to allow you to measure phase all the way through the system. Other switchers will allow you to disable any blanking processor that may be normally in the program line.

To achieve SC/H phasing throughout a system requires fulfillment of three basic considerations:

1. The signal source must have been created with reference to a sync source that conformed to SC/H phasing requirements. (Most sync generator circuits *do* nowadays.)
2. The SC/H phasing must be maintained so that there is no shift whenever the SC/H video signal sources are switched (as would happen in a video production switcher).
3. Any recorders used to record the signal *must* maintain SC/H phasing. (*None* of the heterodyne formats maintain SC/H phasing. Some high-dollar processing amplifiers and time base correctors will reestablish SC/H phasing, when required.)

System Problems

Q: What are Schultz's Laws?

A: Schultz said that Murphy was an optimist.

Q: What general procedure do you use to figure out what is causing problems in a picture?

A: First, decide what portion(s) of the picture is at fault. Does the problem show up only in black-and-white portions of the picture, only in color portions of the picture, in both the black-and-white and color portions of the picture, or only in picture stability and position? To separate color and black-and-white, you can turn down the COLOR, CHROMA or SATURATION control on the monitor.

If the problem exists only in the color portion of the picture, the "chrominance" portion of the video signal has a problem. The problem can be in the incoming video signal or it can be in the picture monitor. Try another picture monitor or look at the signal on a waveform monitor and vectorscope to determine if the monitor or the incoming signal is at fault.

Circuits that will cause chrominance problems have control and signal names that include the terms "chroma;" "C;" "encode;" "decode;" "subcarrier;" "burst;" "balance;" "quadrature;" "red, green or blue pedestal;" "red, green or blue gain;" "saturation;" "hue;" "color;" "tint;" and "3.58."

If the problem exists only in the black-and-white portion of the picture, the "luminance" portion of the video signal has a problem. Again, the problem can be in the incoming video signal or it can be in the picture monitor. Try another picture monitor or look at the signal on a waveform monitor (no vectorscope needed–it displays only chrominance) to determine if the monitor or the incoming signal is at fault.

Circuits that will cause luminance problems have control and signal names that include "luminance," "luma," "Y," "SC Phase," "master pedestal," "brightness," "contrast" and "master gain."

If the problem is with both the color and black-and-white portions of the picture (but not a black bar or problem with position of the picture–discussed next), the problem rests with both the luminance and chrominance parts of the encoded color video signal. The source of this problem lies in the signal path somewhere between the encoder circuit of the camera and the decoder circuit in the picture monitor.

Circuits that will cause problems with both luminance and chrominance have control and signal names that include "video gain," "Y/C gain" and "Y/C delay."

If the problem is the presence of a black bar in the picture (whether moving or stationary) or if the picture moves or jitters, the problem is with the sync portion of the signal. Look at the signal on a waveform monitor. Make sure that the sync is rectangular (no rounded edges) and has an amplitude of 40 IRE units (0.286 Volts) below the video signal blanking level. Another picture monitor *may* or *may not* show you where the source of the problem lies, for the replacement monitor may be more or less sensitive to problems with an incoming sync signal.

Circuits that will cause problems with sync have control and signal names that include "sync," "drive," "deflection," "vertical," "horizontal," "H-phase," "blanking," "hold," "interval," "genlock," "line lock" and "servo."

That's twenty years experience in one-half of a magazine article!

Q: What's a glitch?

A: The completion of a monthly magazine article on time. A manifestation of an aberration, either in a picture or in any other carefully controlled environment.

Q: How can I get rid of hum in a picture?

A: First, what does hum look like in the picture? It normally shows up as thick dark horizontal bands that move upward through the picture. They are the result of 60 Hertz power line frequency signals interfering with the 59.94 Hertz television signals.

There are two common causes of hum in a picture. First, and the less common of the two causes, a component in the filter circuit of a power supply is allowing some of the varying "A.C." voltage to get into the "D.C." power that is necessary for proper operation. With this cause of the problem, a technician needs to determine which component has failed and replace it.

The second cause (and more common with today's technology) comes from the interconnection of unequal electrical "ground" potentials between two pieces of communications equipment. The fix is to equalize the ground potentials. This is often much harder than it sounds.

The easiest "fix" is to interconnect the chassis of the two pieces of equipment with a heavy, very conductive wire. This leads to:

Suggestion #1: Minimize the resistance among various ground potentials. Make sure that there are heavy wires providing very low resistance among the grounds. You really want to try to make a "ground plane" that has equal potential across its entire area.

The second "fix" is to try to make sure that there is limited opportunity for the interfering signal to be developed. This leads to:

Suggestion #2: Minimize the number of possible paths to the unequal ground potentials. This is one of the toughest things to realize.

As a subset of Suggestion #2, you might try an "isolation transformer" in your audio lines *and* a "hum buck coil" in your video lines. A ground loop established in the audio system sometimes induces hum in the picture. (This may even happen without noticing a hum in the sound.)

You can expect a ground loop problem to be experienced when using long cable runs. As a matter of planning, if your cable run is longer than 100 feet or so, have the isolation transformers and hum buck coils standing by. Definitely have some on board a remote production van!!

One more suggestion to eliminate a ground loop–if you can do it, power your equipment with a battery instead of an AC power supply. The ground potential when powered by battery will float and eliminate the ground loop. I know, I know, this isn't always possible, but . . .

DO NOT use a "ground lifter." A ground lifter is one of those cheap three-prong -to- two-prong power adapters you can get at a hardware store. Some people (of course, not you) use these adapters without properly connecting the long ground prong. Although this procedure often breaks the ground loop, IT IS EXTREMELY DANGEROUS NOT TO CONNECT THE LONG GROUND PRONG!!!! There have been several deaths associated with this procedure. As though the prospect of having to deal with someone's death were not enough, wait until you try to explain it to OSHA and the judge. Be safe, don't do it!

Ground loops, grounding systems, and electromagnetic shielding (a kissin' cousin) are a specialty in electronics all unto themselves. It's a tough concept to grasp and this has been a short explanation, so don't feel bad if you try several fixes before one of them takes.

There are very few references that will help you with ground loop problems. To understand the safety considerations, consult the latest version of the ***National Electrical Code*** (***NEC***) or one of the many NEC handbooks. I don't know of any really good reference about the communications aspects of ground loops–does anyone out there???

Q: What are some of the sources of audio hum?

A: There are many potential sources for audio hum, but keep in mind that if you can't hear the hum in real life, the problem is induced into the system. For this to happen, the interfering hum signal must be introduced via electric or magnetic coupling.

Electric coupling can occur via ground loops (see previous articles) or via a bad power supply. Try lifting grounded shield connectors on balanced cables. (You can't really do anything to solve the problem when using unbalanced signals.)

In some instances, the interfering electric source may be radio frequency interference (RFI). If you encounter RFI, either turn off the radio station, move your equipment to another location (where the electric field has less strength), or shield your equipment. Shielding can be provided by encapsulating the equipment with a grounded layer of metallic fiber or metal foil. (Efficient shielding can be *very* expensive.) This can be a particularly sticky problem.

Magnetic coupling can occur between cables or between electromagnets and cables. The magnetic field can come from a monitor, loudspeaker, transformer (or fluorescent light "ballast"), an adjacent power cable, a motor, or anything that has a coil of wire in it.

Q: How can I get rid of the buzz in the sound when my character generator is in the picture?

A: The buzz is a result of a portion of the video signal getting into the sound signal. Usually this occurs when video and audio signals are modulated into one television signal. The resulting signal is called a radio frequency (R.F.) signal. The buzz is usually a result of overloading the receiver.

A character generator signal is a little bit different from most other video signals. The characters have very sharp edges. The signal that represents these sharp edges are more prone to creating the sound buzz than other signals.

To reduce or eliminate the buzz, reduce the maximum signal level of the characters to 90 IRE or below. Unless you're making an Oxydol commercial, you probably won't be able to tell the difference between 90 IRE white and 100 IRE white.

Q: Can a light dimmer cause interference?

A: You betcha! Be sure to use dimmers that advertise low EMI (electromagnetic interference) and low RFI (radio frequency interference) in your facility. Best yet–use remotely controlled dimmer packs.

Q: Is there a standard procedure by which picture quality may be judged?

A: Yes. Recommendation 500-1, "Method for the Subjective Assessment of the Quality of Television Pictures;" Report 313-4, "Assessment of the Quality of Television Pictures;" and Report 405-3, "Subjective Assessment of the Quality of Television Pictures" have been issued by the International Radio Consultative Committee (CCIR) of the International Telecommunications Union (ITU) associated with the United Nations (UN), with which the United States Department of State has asked the Federal Communications Commission (FCC) to coordinate. Big Time! There's probably more reports, recommendations, and organizations involved.

Non-expert observers are preferred. A grading scale where 1=bad (very annoying impairment) and 5=excellent (imperceptible impairment) is used to describe observations. About five real-scene (not test patterns) picture sequences are shown to the observers. The viewing environment is critically controlled, with room illumination low and observers seated from 4-6 times picture height and carefully controlled background around the display.

The referenced CCIR reports go no further than establishing the general testing environment and procedures. Personally, I go several steps further. I look at certain picture attributes and ask the following questions:

STABILITY and GEOMETRY

Is there any motion in any part of the picture that should not be there?
Is there any obvious geometric distortion in any part of the picture where objects are elongated or shortened?
Is there any change in the size of the picture with changing picture brightness or scenes?

LUMINANCE

How accurate is the representation of contrast ratios (relationships between dark and light picture areas)?
How much noise (snow) is in the display?
Is there any smearing from light picture areas into dark picture areas (and *vice versa*)?
Is there objectionable smearing of moving objects in the picture? (There will always be some.)
Is there any ghosting? Look for outlines to the right of picture details?
Is there any blooming or halo around bright picture details that are in front of a dark background?
Is there any interfering signal in the picture? Is the interference moving through the picture or is it stationary?
Is there any darkening of corners of the picture?
Is sharp focus maintained across the entire display?
How objectionable are any visible individual picture elements (pixels)?
How sharp are the edges of the picture details?

COLORIMETRY

How accurate are the color hues?
Are there any variations in color hues?
Is there any dark outline separating different color hues?
How accurate are the color saturations?
How much noise is in saturated colors?

Appendix

Bookshelf

Bookshelf

Over the years, I have reviewed several books and have liked most of them. Here's a few of my favorites:

My favorite truly-technical source is the ***Television Engineering Handbook***, edited by K. Blair Benson and published by McGraw-Hill. This tome covers a lot and, if you truly want to dig into the theory behind equipment, deserves space on your shelf.

A decent source of component and circuit information is ***Television Electronics:Theory and Servicing*** by Milton S. Kiver and Milton Kaufman, published by Van Nostrand Reinhold.

A good source on information on a limited range of topics is ***Electronic Cinematography:Achieving Photographic Control over the Video Image***. This book, by Harry Mathias and Richard Patterson, was published by Wadsworth.

There's an excellent source published by Newman-Smith and written by yours truly (it *has* to be excellent, doesn't it). It's entitled ***Mastering Television Technology:A Cure for the Common Video***.

My hands-down favorite sources of information are out-of-print books by Harold Ennes and published by Howard W. Sams. This is a wonderful series about television broadcasting. There's ***Television Broadcasting: Tape Recording Systems, Television Broadcasting: Camera Chains, Television Systems Maintenance, Digitals in Broadcasting,*** and several others. If you see them in a used book store, *buy them!* If you don't want them, *I do*!

If you're interested in test signals and signal standards, the various rules, regulations and specifications in their original form can't be beat for accurate information. I often dive into Parts 73 and 76 of the FCC Regulations (available from the U.S. Government Printing Office).

I also maintain a selection of EIA (Electronic Industries Association, 2001 Eye St. NW, Washington, D.C. 20006) specifications. Of particular value is RS-170, RS-170A, RS-189A, and RS-330. Also of interest to some people is the "EIA Recommended Practice for Use of a Vertical Interval Reference (VIR) Signal."

There's lots of other standards and applications publications out there. I'll be trying to pass on more recommended references and sources of information as time goes on. Hold on to your pocketbook–it can get expensive!

Target Technology, Inc. (telephone /916/ 639-2102) has a new application note about "Video Terminations–Their Hidden Effect." The application note covers the level-matching problems created when terminations do not have an accurate 75 Ohm value. If you have a problem with unequal or unstable video levels as you proceed through your system, you may want to call Target Technology, Inc. for your free copy.

I just finished reading ***The Audio Engineering Handbook*** that was edited by Blair Benson. Heavy stuff (with all the mathematics to go along), but boy is it good! The book carries ISBN 0-07-004777-4 at a list price of $79.50.

For those of you less faint-of-heart when it comes to math, there's a fairly new series of books from Howard W. Sams that cover audio from a slightly more operational approach. So far, I've found ***Microphone Manual: Design and Application*** by David Miles Huber (ISBN 0-672-22598-0, $29.95), ***Audio Production Techniques for Video*** by David Miles Huber (ISBN 0-672-22518-2, $29.95), ***Modern Recording Techniques*** by Robert E. Runstein and David Miles Huber (ISBN 0-672-22451-8, $24.95), and ***Principles of Digital Audio*** by Ken C. Pohlmann. There appears to be some duplication of material among these books, as far as I've had time to review them, but the information is good.

There's a new series of video tapes about audio that has been released by First Light Video Publishing. The series consists of three 80-minute programs on "Shaping Your Sound with Reverb and Delay," "Shaping Your Sound with Equalizers, Compressors and Gates," and "Shaping Your Sound with Microphones." Hosted by a seasoned audio-type, Tom Lubin, these tapes (optimized for use on VHS Hi-Fi equipment) explores the use of gobs of those knobs and lights that you find on audio gear. I even learned that "organic space" was more than the room I set aside in my garden for the compost heap! If you're creating audio, teaching audio, or listening to audio, buy these tapes–it's the type of learning experience in sound that you've been looking for all these years! The professional versions, at $119.00 each, provide a learning experience that is like having your very own personal mentor with a full range of audio equipment at his disposal. First Light Video Publishing can be reached at (213) 467-1700.

I just got my greedy little hands on a copy of a new application note from Tektronix entitled ***Television Measurements: NTSC Systems*** (Publication Number 25W-7049 9/89) by Margaret Craig. It's a slick 80 page spiral-bound notebook crammed with absolutely everything you ever wanted to know about making measurements (and then some). I've just had an opportunity to scan the work, but it appears to have plenty of basic stuff thoroughly mixed in with the nitty gritty technical stuff. The appendices cover details of some of the specifications where I go for answers. Some of the measurements described in the application note are not a pretty sight (like "Incidental Carrier Phase Modulation" and "Group Delay with Sin X/X"), but for you technocrats, it's a great reference. Contact Tektronix for your free copy.

The only downside comment I can make about ***Television Measurements*** is the same as the vast majority of test and measurement application notes–it does not directly relate measurement results with subjective judgement of picture impairments. Instead, the reader is referred to more specifications and standards that are difficult for many users to interpret.

As a writer, I *know* that attempting to match numbers with subjective interpretations of distortions is extremely difficult; however, the subjective interpretation by the viewer is the *required* end-result in actual use of television. Everything else is technical doublespeak for the vast majority of users.

Tektronix also has published ***Television Measurements: PAL Systems*** if you want to go continental.

For those of you who are interested in such things, I just picked up a copy of ***Video Scrambling & Descrambling for Satellite & Cable TV*** by Rudolf F. Graf and William Sheets (Howard W. Sams, 1987, ISBN 0-672-22499-2, $19.95). The 250 page book discusses the majority of the scrambling techniques used today, including the VideoCipher II® and B-MAC systems (if you want to consider B-MAC an encryption system.

Although several obligatory cautions are plastered throughout the book about pirating signals without appropriate compensation, several construction projects for descramblers are presented. These projects are of descramblers of *basic* systems, not the complex commercial systems.

The book is OK, but the consistent use of the spelling of "synch" instead of the industry-standard "sync" really made me question the validity of some of the discussions. "Synch" appears so many times that I eventually gave up trying to read the book. (Maybe I'm out-of-line about "synch," but when somebody tries to communicate a difficult concept, why invent a nonsensical change the mode of communication.)

The book is written for technocrats, not for producers. If you want or need technical information about video signal scrambling techniques, get the book. If you don't need the information, move on–I don't think you're missing anything.

Here's another book that belongs on your bookshelf if you're serious about television and audio technology. ***Television and Audio Handbook for Technicians and Engineers*** is a collection of specialized chapters written by some of the best in their field. K. Blair Benson and Jerry Whitaker collaborated to develop this work. It appears to be an updated continuation of the ***Television Engineering Handbook*** and the ***Audio Engineering Handbook***, both by Blair Benson. Take a close look at this one–it's a winner (McGraw-Hill, 1990, ISBN 0-07-004787-1, $39.95).

There's a new book out that should be of interest to those of you who design systems that try to interface computers with television equipment. ***Digital Video in the PC Environment*** by Arch C. Luther (Intertext Publications, McGraw-Hill, 1989, $27.95, paper) attempts to describe the process of getting a computer to talk the language of television.

The book begins with a discussion of conventional analog television technology and an overview of television standards. Personally, I felt like I was expected to have known the majority of the information here and the chapter was simply a review for those who already know the details. The chapters on digital video and audio that follow form a basis for understanding the minimum processing and operational capabilities that are imposed on computers that will fulfill the requirements of television systems. Later chapters cover digital still video image systems, digital picture manipulation (including a sample program written in C for a VAX computer), and details of the Digital Video Interface (DVI) being developed by the computer industry.

If you are working to develop computer applications in the video industry or if you're interested in the details of such stuff, Digital Video in the PC Environment belongs on your bookshelf. The information you need is probably in there, although it may take some digging to get it out.

I found another new book that I really like. ***The Art of Digital Video*** by John Watkinson (ISBN 0-240-51287-1, Focal Press, 1990, $49.95) is the best introduction to the world of digitized video that I have yet seen. This is a book for the advanced technician that wants to know more about the theory behind digital video operations, as well as specific applied technologies.

The Art of Digital Video has chapters on conversion schemes between the digital and analog domains, digital video signal coding and processing, digital magnetic and optical recording (including the D-1 and D-2 tape formats), and the interconnection of digital video devices. Of particular interest to me is the chapter about the use of disk drives in the digital video environment–this, to me, is the "state-of-the-arc" and we'll soon see wider use of this technology. In addition to the coverage of digital video, digital audio concepts are presented for those applications where audio is integrated into the video environment.

The book, published in England, is geared toward the PAL production environment. Don't let this deter you, for there is also comprehensive coverage of NTSC and SECAM environments. The approach is interesting and, if you keep your eyes open, there's humor imbedded in some of the discussions.

If you have some technical background and you want to get up-to-speed on the digital environment ASAP, get a copy of ***The Art of Digital Video***. It's so good, I wish I had written it!

Normally, I don't venture too far from the field of technical television, but I finally found a book on desktop publishing that I really liked. ***The Professional Look: The Complete Guide to Desktop Publishing*** by Stephen E. Manousos and Scott W. Tilden (Venture Perspectives Press, San Jose, ISBN 0-932309-40-2, $19.95) is one of those rare books that give a good, realistic overview of a technically complex subject. If you're interested, involved, or infatuated with computer-aided publishing, take a look at this one. It's a winner.

Bibliography

Bartlet, George W. (ed). *National Association of Broadcasters Engineering Handbook.* 6th ed. Washington, D.C.: National Association of Broadcasters, 1975.

Bensinger, Charles. *The Video Guide* 2nd ed. rev. Santa Fe, New Mexico: Video-Info Publications, 1981.

Beynon, J.D.E. and Lamb, D.R. (ed.). *Charge-coupled devices and their applications.* St. Louis: McGraw-Hill Book Company (UK) Limited, 1980.

Benson, K. Blair (ed. in chief). *Television Engineering Handbook.* New York: McGraw-Hill Book Company, Inc., 1986.

Cartwright, Steve R. *Training with Video.* White Plains, New York: Knowledge Industry Publications, Inc., 1986.

Castleman, Kenneth R. *Digital Image Processing.* Englewood Cliffs, New Jersey: Prentice-Hall, Inc., 1979.

Crutchfield, E. B. (ed.). *National Association of Broadcasters Engineering Handbook.* 7th ed. Washington, D.C.: National Association of Broadcasters, 1985.

Cunningham, John E. *Cable Television.* 2nd ed. Indianapolis: Howard W. Sams & Co., Inc. 1980.

Ennes, Harold E. *Digitals in Broadcasting.* Indianapolis: Howard W. Sams & Co., Inc., 1977.

Ennes, Harold E. *Television Broadcasting: Equipment, Systems Systems, Operating Fundamentals.* 2d ed. Indianapolis: Howard W. Sams & Co., Inc., 1979.

Ennes, Harold E. *Television Broadcasting: Tape Recording Systems.* 2nd ed. Indianapolis: Howard W. Sams & Co., Inc., 1979.

Ennes, Harold E. *Television Broadcasting: Systems Maintenance.* 2nd ed. Indianapolis: Howard W. Sams & Co., Inc., 1978.

Fredida, Sam and Malik, Rex. *The Viewdata Revolution.* London: Associated Business Press, 1979.

GTE Sylvania. *Lighting Handbook for Television, Theatre, Professional Photography.* 6th ed. Danvers, Massachusetts: GTE Sylvania Incorporated; Lighting Center, 1978.

Giloi, Wolfgang K. *Interactive Computer Graphics: Data Structures, Algorithm, Languages.* Englewood Cliffs, New Jersey: Prentice-Hall, Inc. 1978.

Hansen, Gerald L. *Introduction to Solid-State Television Systems: Color and Black & White.* Englewood Cliffs, New Jersey: Prentice-Hall, Inc., 1969.

Harris, Cyril M. (ed). *Handbook of Noise Control.* (2d ed). New York: McGraw-Hill Book Company, 1979.

Hutson, Geoffrey H. *Colour Television Theory: PAL-System Principles and Receiver Circuitry.* St. Louis: McGraw-Hill Publishing Company (UK) Limited, 1971.

Kiver, Milton S. and Kaufman, Milton. *Television Electronics: Theory and Servicing.* 8th ed. New York: Van Nostrand Reinhold Company, Inc., 1983.

Kybett, Harry. *Video Tape Recorders.* 2d ed. Indianapolis Howard W. Sams & Co., Inc., 1978.

Lancaster, Don. *TV Typewriter Cookbook.* Indianapolis: Howard W. Sams & Co., Inc., 1976.

LeTourneau, Tom. *Lighting Techniques for Video Production: The Art of Casting Shadows.* White Plains, New York: Knowledge Industry Publications, Inc., 1987.

Luther, Arch C. *Digital Video in the PC Environment: Featuring DVI Technology.* New York: Intertext Publications/McGraw Hill, 1989.

Marsh, Ken. *Independent Video: A Complete Guide to the Physics, Operation, and Application of the*

New Television for the Student, the Artist, and for Community TV. New York: Simon and Schuster, 1974.

Martin, James. *Telematic Society: A Challenge for Tomorrow.* Englewood Cliffs, New Jersey: Prentice-Hall, Inc., 1981.

Mathias, Harry and Patterson, Richard. *Electronic Cinematography.* Belmont, California: Wadsworth Publishing Company, Inc., 1985.

McGinty, Gerald P. *Video Cameras: Operation and Servicing.* Indianapolis: Howard W. Sams & Company, Inc., 1984.

McQullin, Lon. *The Video Production Guide.* Indianapolis: Howard W. Sams & Co., Inc., 1983.

Mee, C. Denis and Daniel, Eric D. (ed). *Magnetic Recording Handbook: Technology and Applications.* New York: McGraw-Hill, 1990.

National Fire Protection Association. *National Electrical Code 1984 Edition.* Quincy, Massachusetts: National Fire Protection Association, 1983.

Paulson, C. Robert. (Principal Author). *BM/E's ENG/EFP/EPP Handbook: Guide to Using Mini Video Equipment.* New York: Broadband Information Services, Inc., 1981.

Prentiss, Stan. *HDTV: High Definition Television.* Blue Ridge Summit, PA: Tab Books, 1990.

Remley, Frederick M. (ed.). *One-Inch Helical Video Recording.* Scarsdale, N.Y.: Society of Motion Picture and Television Engineers, 1978.

Robinson, Richard. *The Video Primer: Equipment, Production, and Concepts.* New York: Quick Fox Books, 1974.

Sigel, Efrem, Schubin, Mark, and Merrill, Paul F. *Video Discs: The Technology, the Applications and the Future.* New York: Van Nostrand Reinhold Company, 1980.

Sigel, Efrem (ed.). *The Coming Revolution in Home/Office Information Retrieval.* New York: Harmony Books, 1980.

Smith, Coleman Cecil. *Mastering Television Technology: A Cure for the Common Video.* Richardson, Texas: Newman-Smith Publishing Co., Inc., 1988.

Smith, Coleman Cecil. *Instruction Manual for Mastering Television Technology: A Cure for the Common Video.* Richardson, Texas: Newman-Smith Publishing Co., Inc., 1988.

Smith, Coleman Cecil. *Experiments to Master Television Technology.* Richardson, Texas: Newman-Smith Publishing Co., Inc., 1991.

Tremaine, Howard M. *Audio Cyclopedia* 2d ed. Indianapolis: Howard W. Sams & Co., Inc., 1973.

Tremaine, Howard M. *Passive Audio Network Design.* Indianapolis: Howard W. Sams & Co., Inc., 1964.

Ulz, Peter. *Video User's Handbook.* Englewood Cliffs, New Jersey: Prentice-Hall, Inc., 1980.

Zetl, Herbert. *Television Production Handbook.* 4th ed. Belmont, California: Wadsworth Publishing Company, Inc., 1984.

Watkinson, John. *The Art of Digital Video.* Boston: Focal Press, 1990.

Index

Symbols

A

B

C

D

G

H

I

J

L

M

N

O

P

Q

R

T